CAPITALISM AND ARITHMETIC

✠CAPITALISM AND ARITHMETIC

✠ The New Math of the 15th Century ✠ Including the full text of the *Treviso Arithmetic* of 1478 ✠ Translated by David Eugene Smith

✠FRANK J. SWETZ

✠Open Court
La Salle, Illinois

First printing 1987.
Second printing 1989.

Printed and bound in the United States of America.

Library of Congress Cataloging-in-Publication Data

Swetz, Frank.
Capitalism and arithmetic.

Bibliography: p.
Includes index.
1. Mathematics—History—15th century. 2. Arithmetic—History—15th century. I. Smith, David Eugene, 1860–1944. II. Title.
QA23.S94 1986 510'.9'024 85-21694
ISBN 0-87548-438-7
ISBN 0-8126-9014-1 (pbk.)

Dedicated to my parents,
Frank Leo Swetz
and
Helen Curtis Swetz,
who taught me how to
persevere.

This book includes the first English translation of the *Treviso Arithmetic* of 1478, the earliest known example of a printed book on arithmetic. The *Treviso Arithmetic* bore no title and no author's name; it appears in many library catalogues listed as *Arte del Abbaco*. The present Open Court edition was designed by John Grandits, and is set in Bembo, an English Monotype design based on a face cut by Francesco Griffo for the printer Aldus Manutius of Venice in the early 1490s. The small Maltese Cross motif was a feature of the original Treviso volume.

Calligraphy is by Joan Armstrong. Open Court thanks Kurt Schuler for expert editorial advice. Thanks also to Irving Adler for a helpful suggestion which improved upon the first printing.

MASTER: Wherefore in all great works are Clerks so much desired? Wherefore are Auditors so well fed? What causeth Geometricians so highly to be enhaunsed? Why are Astronomers so greatly advanced? Because that by number such things they finde, which else would farre excell mans minde.

SCHOLAR: Verily, sir, if it bee so, that these men by numbering, their cunning do attain, at whose great works most men do wonder, then I see well I was much deceived, and numbering is a more cunning thing than I took it to be.

Robert Recorde, *The Declaration of the Profit of Arithmeticke* (1540)

Contents

Preface

Several years ago, while searching through miscellaneous papers and documents of the David Eugene Smith Collection of the Butler Library at Columbia University, I came across a time-yellowed manuscript. The dusty bundle of pages bore the simple title *The Treviso Arithmetic 1478*, and consisted of over two hundred pages of typed and handwritten notes. In the main, the work was a translation Smith had made of an Italian arithmetic book with the intent of publishing it. In the nineteen twenties, David Eugene Smith was an extremely active scholar in the fields of mathematics education and the history of mathematics.* Chairman of the mathematics department at Teachers College, Columbia University, and a prolific author of textbooks, Smith's varied demands distracted him from

*See the brief historical sketch of the life and work of David Eugene Smith in the epilogue following the text.

his *Treviso Arithmetic* project, and the work was never completely prepared for publication.

In glancing through the manuscript, I found that it dealt with the concepts and processes of early Renaissance arithmetic, involving the use of Hindu-Arabic numerals, their computational algorithms, and the use of those algorithms in solving problems of commercial arithmetic. Its problem situations focused on concerns of Italian merchants of the late fifteenth century, such as payment for goods received, currency exchange, and the determination of shares of profit derived through partnership arrangements. While the contents were interesting, my main concern at the time focused on other matters, and I continued my search, dismissing the Treviso manuscript for the moment.

In the ensuing months after my initial encounter with Smith's papers, my thoughts repeatedly returned to the Treviso translation. Gradually, its importance and significance dawned upon me. The fifteenth century was a time of commercial, intellectual, and mathematical ferment, yet there is a dearth of specific information on the mathematical climate of this period. Some work has been done in this area, notably by Gino Arrighi of Italy, Kurt Vogel of Germany, and, most recently, by Warren Van Egmond of the United States. Arrighi, perhaps the most active contemporary researcher on the history of mathematics in medieval and early Renaissance Italy, has produced a rather extensive collection of research publications that reflect on this subject. Vogel, of the Deutsches Museum, Munich, has published two major works that contribute to an understanding of the mathematical activity of this time: *Die Practica des Algorismus Ratisbonensis: Ein Rechenbuch des Benediktiner klosters St. Emmeran aus der Mitte des 15 Jahrhunderts* (1954), and *Ein italienisches Rechenbuch aus dem 14 Jahrhundert* (1977). However, the works of both these men are limited to readers of their respective native languages, Italian and German, and are thus not readily available to a wide transatlantic audience. Dr. Van Egmond's scholarly study, *The Commercial Revolution and the Beginnings of Western Mathematics in Renaissance Florence, 1300–1500* (1976), focuses on the broad impact of "abaci" manuscripts on mathematical thinking, and attempts to view their contents and development within an economic and sociological framework. Unfortunately, its format and availability as a doctoral dissertation limits its accessibility and appeal to a general reading audience. Thus, I began to realize that Smith's work was the first and only English language translation of a fifteenth-century European arithmetic book that I had encountered. Its contents, if analyzed, could provide valuable information. A little further research revealed that the Treviso manuscript was, indeed, an intellectual and historical treasure for several reasons:

1. It is the earliest known dated, printed arithmetic book.
2. It is one of the first mathematics books written for popular consumption and, as such, marks a turning point in the history of human knowledge. The book's message, transcribed in the common Venetian dialect of the period, was intended for all who wished to learn

the art of computation, not just for a privileged few, as had been the case previously.

3. The contents of the work provide an early example of efforts to promote the Hindu-Arabic numeral system and its computational algorithms. Use of the abacus and Roman numerals was still popular in much of Europe, although to a lesser degree in Italy, where commercial interests demanded more efficient computational procedures. Thus, the translation supplies insights into the mathematical climate and controversy of the late fifteenth century.

More fully aware of the importance of the manuscript I had uncovered, I undertook to complete the task that David Eugene Smith had begun nearly eighty years before—to publish a study of early Renaissance arithmetic based on the contents of the 1478 Treviso tome. Smith's concern was mainly pedagogical; he was primarily interested in how early arithmetic was taught. The following study, while considering the heuristics of early European arithmetic instruction, will also investigate the mathematical and sociological significance of the *Treviso Arithmetic*'s contents.

The first chapter of this study provides a general background in which the *Treviso Arithmetic* and its contents must be understood. Chapter Two presents Smith's free translation of the *Arithmetic*. The succeeding four chapters serve as a commentary on its contents and correspond to the sequential ordering of topics in the *Arithmetic:* numeration, addition and subtraction, multiplication, and division, and the techniques of problem solving. Chapter

Seven focuses on the social, economic and commercial aspects of the book's contents, and summarizes the findings of this excursion into early Renaissance mathematics.

Ostensibly, this study focuses on a book and its contents but, perhaps more importantly, it also concerns a time (the early Renaissance), a place (the Venetian Republic), and circumstances (the rise of mercantile capitalism and the economic beginnings of industrialization), and how these three aspects molded and affected the directions of human involvement with mathematics. Ultimately, the object of this work is to contribute to a better understanding of the historical interaction of the development of mathematical ideas and techniques and their societal milieu.

The undertaking and completion of a study of this magnitude and scope, relied on the cooperation and assistance of many individuals and institutions. I would like to acknowledge and thank some of the principal participants who assisted me in my research tasks. The libraries at Columbia University kindly provided me with access to their copy of the *Treviso Arithmetic* housed in the George A. Plimpton Collection, Rare Book and Manuscript Library, and to the notes and translation of D. E. Smith contained in the David Eugene Smith Collection. My own library staff at the Heindel Library, Capitol Campus, the Pennsylvania State University, particularly Carolyn Miller, reference librarian, and Ruth Runion, interlibrary loans clerk, demonstrated resourcefulness and determination in securing elusive references for my use. An uncle, Arthur Torelli, assisted in the translation of Italian language materials.

Dr. Alan M. Stahl, associate curator of medieval coins, the American Numismatic Society, supplied information about early European currency. Stillman Drake, of the Institute for the History and Philosophy of Science and Technology, University of Toronto, clarified some issues on Renaissance timekeeping. Finally, I wish to thank Dr. Peter Hinton for his valuable editorial suggestions, and my wife, Joan, for her careful proofreading of the manuscript.

Frank J. Swetz
Harrisburg, PA
1986

CAPITALISM AND ARITHMETIC

✠ 1
Perspectives

The Climate of European Social and Intellectual Change

Fifteenth-century Europe witnessed a period of great change—social, economic, and intellectual—and was marked by both the physical and transcendent movements of peoples and ideas. The historical rise and coming together of the forces and conditions that bred these changes is a complex phenomenon beyond the scope of this study; however, the changes or movements themselves are worth noting (see Table 1.1). Feudalism was on the wane. The beginnings of a spirit of national or regional consciousness were emerging among peoples who began to feel a sense of societal cohesiveness and purpose. Growth of towns and cities meant a new distribution of political power, as well as new types of power. Occupational opportunities broadened from the realms of

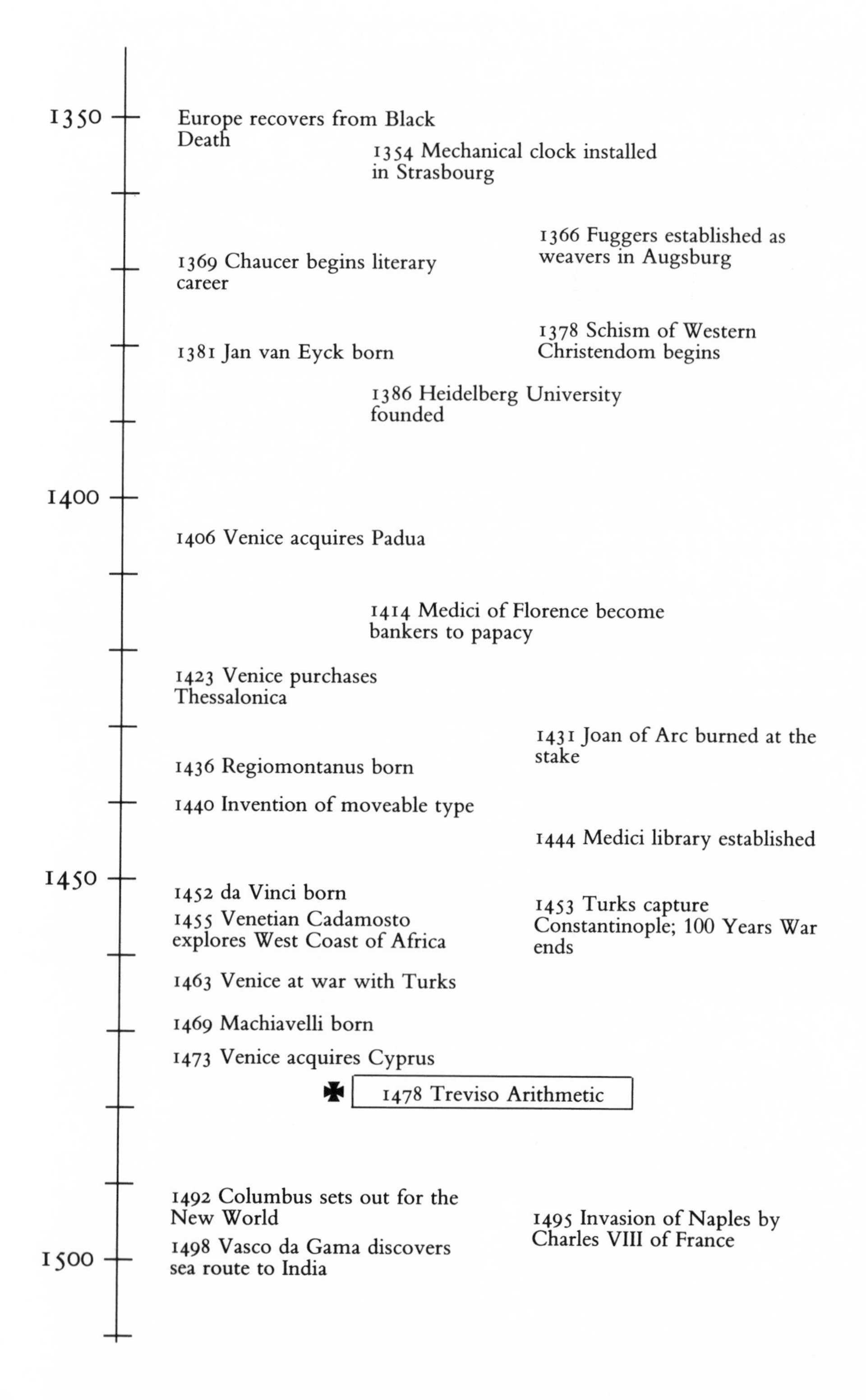

Table 1.1 A Chronology of Events in the Time of the *Treviso Arithmetic*

1500

1509 Reign of Henry VIII in England begins

1517 Protestant Reformation begins

1522 Magellan completes circumnavigation of the world

1527 Armies of Charles V sack Rome

1535 Thomas More executed

1542 Universal Inquisition established by Pope Paul III

1543 Copernicus publishes *De Revolutionibus*

1545 Council of Trent convened

1547 Spanish Inquisition

1550

1558 Elizabeth ascends throne in England

1561 Francis Bacon born

1565 St. Augustine, Florida established as first settlement in North America

1569 Gerhardus Mercator publishes his projection map of the world

1582 Gregorian calendar established

1585 Simon Steven publishes *La Disme*

1586 Montaigne essays published

1588 Defeat of Spanish Armada

1590 Microscope invented

1592 Galileo debates principles of gravity

1595 Dutch East India Company establishes claims in Orient

1596 Descartes born

1600

1600 British East India Company challenges Dutch for control of spice trade; first opera, Florence

1614 Napier introduces logarithms

1616 Shakespeare died

1622 Molière born

1636 Harvard College founded

1642 Newton born

1650

agricultural and pastoral pursuits to include participation in activities of manufacture and commerce. One way this new power manifested itself was through entrepreneurship in trade and manufacture. Commercialism spawned wealth that was re-invested, and modern mercantile capitalism was born.[1] Gradually a new aristocracy arose, one whose status was determined by the accumulation of wealth rather than a blood inheritance. In the climate of relative social stability that could be secured through wealth, man could, once again, afford to examine the institutions and world around him.[2] At first, this exploration was intellectual. Ideas and concepts, many of which were previously held sacrosanct, were re-examined, questioned, and held up to a new scrutiny. Initially, answers to many of these questions were sought in the classics of antiquity, but gradually a new rationalism based on sense experience and systemized investigation was found to supply more satisfying answers. A world view supported by empirical fact began to replace institutions and beliefs based solely on traditional faith and a sense of spiritual integrity. The appearance of new power bases, combined with the growth of intellectual curiosity, resulted in a rise of secularism, weakening the central authority of the Catholic Church. Both authority and tradition lost their former weights. Buoyed up by an acquired sense of self-confidence, both individuals and territorial regions sought more firmly to control their own destinies. A new intellectual and creative energy was evident, and it manifested itself in a flowering of human expression, particularly in the arts, literature, music, and architecture. A revival, a rebirth of the human spirit, was taking place, and the hallmark of this historical period became the word *Renaissance,* a term coined with the advantage of historical retrospection.[3]

While the combination of these movements resulted in the macro-historical changes of the time, particular causal relationships, chain reactions if you like, can also be discerned. A spirit of nationalism fostered social unity, allowing inter-regional commerce to flourish, which in turn perpetuated capitalism. Intellectual humanism was patronized by capitalism and secularism, which broadened man's horizons of inquiry and innovation. This chain of events affected every aspect of human life. While, most notably, the arts flourished, the orientation of science and scientific thinking also began to change. Available documents, manuscripts, and books substantiate this fact. One such piece of evidence, a mathematics textbook known simply as the *Treviso Arithmetic,* provides us with a fascinating glimpse of the mathematical transition taking place, and hints at some of the forces shaping it. The book bears no formal title and its author remains anonymous; therefore, it has been named after its place of origin. Treviso is a northern Italian city, and a question that immediately comes to mind is why an arithmetic book was published in a relatively obscure place rather than in an acknowledged center of scholarly activity such as Paris or Rome.

The Venetian Republic and Treviso

As early as the tenth century, the city of Venice was beginning to assert its authority on the European scene as an important mercantile power. Its location on the Adriatic Sea, with direct access to the Mediterranean and the Levant, played a decisive role in its securing a commercial and maritime status; but an even more important factor in its rise to economic supremacy was the spirit of its people. The Venetians, historically schooled in the strategies of survival in which adversity is turned into advantage,[4] readily used their powers of innovation and natural aggressiveness to seize upon the financial opportunities provided by the revival of commerce among their northern neighbors. The merchants of Venice soon became renowned for their business and financial acumen, and the city itself became an international entrepôt, and the first truly capitalist center of Europe. A sense of adventure was in the air. Martino da Canale (c. 1267), an observer of the time, noted, "merchandise passes through this noble city [Venice] as water flows through fountains".[5] Fortunes were to be made or lost in the process.

The political climate of Venice was also conducive to a spirit of enterprise. Venice and its eventual holdings were incorporated into a republic ruled by an elected doge, or duke, and his advisory councils, the most powerful of which was the Grand Council, composed of elected nobles. This patrician regime provided stability and an aura of justice for its charges, and kept the Republic independent of foreign domination. A strong sense of political identity and pride was fostered among the citizens of Venice. Without the constraints imposed by despotic feudal overlords, individual independence prevailed, and opportunities for personal initiative flourished. Venetian independence also extended to its relationship with Rome and the Catholic Church. While Venetians remained respectful of the Church, its institutional power in Venice was legislatively controlled. Thus, the Venetian Republic had a firm hold on its own fate and course of progress.

By the fifteenth century, Venice's commercial ventures had made it the trade capital of Europe, and one of the richest cities of the known world. The rule of the Republic now encompassed two empires: the first consisted of a string of colonies and naval bases extending down the Adriatic Coast around the Balkan peninsula and into the eastern Mediterranean; the other included Venice and stretched westward across the Lombard plain to the borders of the Duchy of Milan, and enclosed such northern Italian cities as Revenna, Treviso, Padua, Vicenza, Verona, and Brescia.[6] Venetian commercial skill became the driving force in a system of exchange which conveyed silks, furs, and spices from Turkestan to the Baltic, and from Cathay to the North Sea. Ghent, Bruges, Antwerp, Amsterdam, London, and cities of the Hanseatic League were rich clients ready to do business with the merchants of the south. A myriad of goods, mundane and exotic, passed through the warehouses and toll stations of Venice. While exerting almost a total monopoly on the Eastern spice trade, the Venetians also controlled the import of cotton into Europe from Cyprus, Palestine, and Syria. They were the sole suppliers for Europe's developing fustian industry.[7] The weavers of Augsburg, Ulm, and other German cities depended on Venice both for their raw cotton supply and eventual accessibility of their products to eastern markets. Venice also became a transit point for wool and developed its own wool processing, dyeing, and weaving industries.[8] Oriental industries such as paper making and silk weaving were transplanted to Venice and helped to broaden and further strengthen its economic base. Precious metals, especially gold and copper from Hungary and the Tyrol, passed through Venice on their way eastward. Soon Venice also found itself the European center for the processing, transportation, and coining of these metals.[9]

In its expansion westward, the Venetian Republic acquired Treviso in 1339. The city lies on the fertile agricultural plains of Veneto, about 26 kilometers northwest of Venice, a day's travel by the standards of the Middle Ages, and is situated at the confluence of the Sile and Cagriairo rivers.[10] Treviso's location on the main northern trade route to Vienna and the German cities accessible through the Brenner Pass, as well as the main trade route to the Lombard hinterlands, made it an extremely valuable commercial link in the republic's trade empire. During the plague years, the city served as a refuge for Venetians (See Figure 1.1). In the fifteenth century, Treviso had a population of about 12,000, compared with Venice's 150,000.[11] While it was a small city,

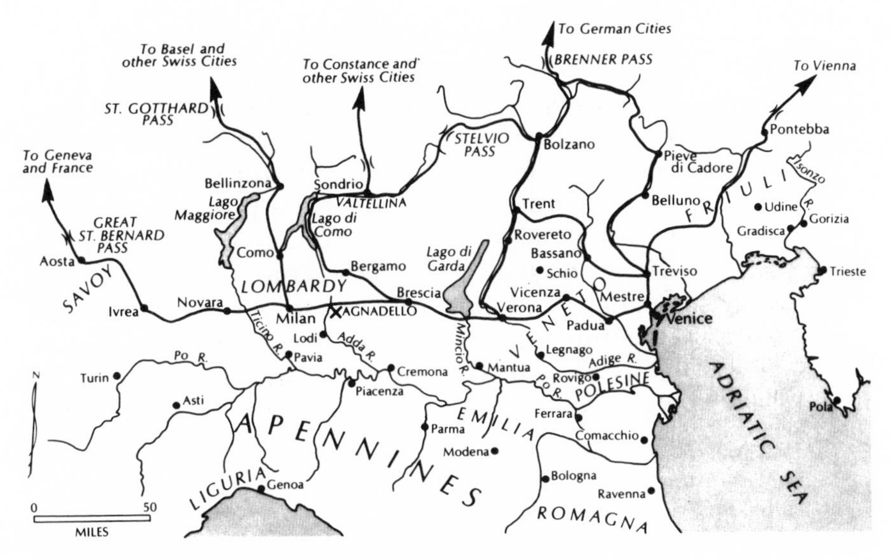

Figure 1.1 Trade routes passing through Treviso.

and a political and economic satellite of Venice, it still had a viable industrial and commercial identity of its own. Venetian merchants, in aspiring to the position of landed gentry, acquired mainland estates. Many of these estates and country houses were in the area of Treviso.[12] Such estates and farms around Treviso supplied Venice with grains and other foodstuffs for consumption. The region's abundant fresh water supply nurtured several industries, including papermaking, boat construction, and the milling of rice. Venice sent its woolen cloth to Treviso to be fulled, washed, and softened.[13] At various times, it also supported a mint (c. 773)[14] and thriving cement, ceramic, and textile industries.

Commercial arithmetics of earlier days make mention of this city in connection with trade to Venice, Padua, and other cities of the region, and it is known that a practitioner of commercial arithmetic was active in Treviso in the year 1372.[15] Thus, the commercial and financial activities of Treviso at the close of the fifteenth century would seem to warrant the skills of trained computers, and the existence of books on the subject of commercial arithmetic.

Welsche Praktik

By the beginning of the sixteenth century, there was a lag of perhaps two hundred years between the commercial and financial methods of the Hanseatic League merchants and their more advanced southern counterparts. From the fourteenth century on, merchants from the north travelled to Italy, particularly to Venice, to learn the *arte dela mercadanta,* the mercantile art, of the Italians. Sons of German businessmen flocked to Venice to study the *Welsche Praktik,* the foreign practices of business, commercial arithmetic and currency exchange. After acquiring these skills, they returned home with a new Italian vocabulary which included such terms as: *disagio,* discount; *credito,* credit; *valuta,* value; *netto,* at a net price, etc. [16] Most of these words are evident in the jargon of the contemporary business world. Even Jacob Fugger, the German merchant prince, left Augsburg to study business techniques in Venice. In fact, the flow of German merchants to Venice seeking either knowledge or business was so great that a special complex was established to accommodate the visiting Germans. The *Fondaco dei Tedeschi* or German Factory in Venice, a five-story structure with a great courtyard, provided accommodations for over 80 visiting merchants and their servants and served as a combination warehouse, hotel, and market.[17] The painter Titian served his apprenticeship frescoing its walls. While the spectrum of sought-after knowledge was broad and included the whole sum of practical business models used by the Italians in the areas of exchange, wholesale trade, manufacturing, colonial administration, banking, and high finance, a fundamental skill all visiting merchants hoped to acquire was a proficiency in the methods of Italian commercial arithmetic. In many instances this *Welsche Praktik* became the singular reason for a migration southward.

Early, during their rise to commercial supremacy, the Italians, and particularly the Venetians, realized the importance of the use of arithmetic in their daily business transactions. In trade contacts around the Mediterranean and Barbary coasts, Italian merchants became exposed to the Hindu-Arabic numeral system and its methods of computation. Raised in a Pisan trading colony in Bugia (Bougie), in what is now Algeria, Leonardo of Pisa (1180–1250) studied this new system of arithmetic under the guidance of an Arab master.[18] He became convinced that the new numerals and their methods were vastly superior to the Roman numerals commonly employed in Europe. Leonardo, also known as Fibonacci (the son of Bonaccio), became the evangelist of the new knowledge and published his impressions in a book, *Liber abaci* (1202).[19] Much of the text was a general introduction to the Hindu-Arabic numerals and their algorithms; however, a special section reflected Fibonacci's professional background and considered commercial applications of arithmetic. The book and its message were well-received in the *fondacos,* or merchant houses, of Pisa, Genoa, and Venice, and soon the Hindu-Arabic symbols were replacing Roman numerals in account books, and the use of the abacus was giving way to computations performed with pen and ink. This new mathematical knowledge was also being introduced into Europe via translations of Arabic works emanating from Spain, but the receptive climate in Italy soon gave the Italians a predominance in this art. The Italians not only adapted new mathematical techniques for their commercial interests, but also became innovators—the practice of double entry bookkeeping originated in northern Italy during this period. In the technique of double entry bookkeeping, each entry is recorded twice, on one page under a debit heading and on the opposite page under a credit title. This method of financial recording was employed in several Italian cities by the beginning of the fourteenth century, and by the year 1400 it had been well established in Italy as a basic tool of financial management.[20] The most complete early guide to double entry bookkeeping was included in Luca Pacioli's *Summa de arithmetica* (1494), in which he referred to the process as *scrittura alla Veneziana,* in the written manner of Venice.[21] Indeed, the Venetian refinements of the art of accounting were openly acknowledged, admired, and copied by visiting foreign merchants.[22] The Venetians continued to strengthen their position in commercial and commercially-related mathematics by being the first city in Italy to create a university chair of mathematics devoted to navigation.[23] Piero de Versi, an early holder of this chair, produced one of the first tracts combining both the arts of mathematics and navigation, *Alcune raxion dei marineri* (1444). Venice eventually went on to become the first European city to endow public lectures on algebra.

The climate of mathematics and mathematics education in Italy was of such excellence that it attracted students from all over Europe. An anecdote related by Tobias Dantzig describes the situation:

> There is a story of a German merchant of the fifteenth century, which I have not succeeded in authenticating, but it is so characteristic of the situation then existing that I cannot resist the temptation of telling it. It appears that the merchant had a son whom he desired to give an advanced commercial education. He appealed to a prominent professor of a university for advice as to where he should send his son. The reply was that if the mathematical curriculum of the young man was to be confined to adding and subtracting, he perhaps could obtain the instruction in a German university; but the art of

> multiplying and dividing, he continued, had been greatly developed in Italy which, in his opinion, was the only country where such advanced instruction could be obtained.[24]

Indeed, this was true. Italy was the place to learn computational mathematics, but, even in Italy, if one wished to study numerical computing, particularly that associated with commercial endeavors, it was not necessarily at the university that one studied. By the middle of the thirteenth century, arithmetic was being taught as a science at European universities. Although instruction was eventually provided in the methods of the Hindu-Arabic numeration system, the level of teaching was often theoretical and devoid of practical applications.[25] University education was classical in nature, and even though arithmetic was one of the subjects of the *quadrivium,* under the influence of scholasticism its usefulness was seriously questioned in comparison with its companion subjects, particularly grammar, logic, and rhetoric, the *trivium.* In Henri d 'Andeli's *Battle of the Seven Arts,* a commentary on the classical curriculum of medieval universities, the author depicts arithmetic as a rather aloof maiden detached from the reality surrounding her:

> Arithmetic sat in the shade,
> Where she says, and she figures,
> That ten and two and one make thirteen,
> And three more make sixteen;
> Four and three and nine to boot
> Again make sixteen in their way;
> Thirteen and twenty-seven make forty;
> and three times twenty by themselves
> make sixty;

Five twenties make a hundred and ten
hundreds a thousand.
Does counting involve anything further?
No.
One can easily count a thousand thousands
In the foregoing manner.
From the number which increases and
diminishes
And which in counting goes from one to
hundred.
The dame makes from this her tale,
That usurer, prince and count
Today love the countress better
Than the chanting of High Mass.[26]

In universities where practical arithmetic was taught in conjunction with theoretical arithmetic, the fees for studying theoretical arithmetic were often twice the amount charged for practical work, demonstrating the perceived discrepancy between the value of the two subjects. If a student in the early Renaissance wished to learn commercial mathematics, he usually did not go to the university, but sought out a reckoning master, a man skilled in the arts of commercial computation, with whom to study.

Europe's mercantile development from the thirteenth century onward placed increased importance on an understanding of, and proficiency in, commercial arithmetic (see Figure 1.2). Merchants found themselves obliged to instruct their sons and apprentices in the mathematical arts they knew. As the tempo of trade increased under the stimulation of the Hanseatic League's activities, merchants or guilds had to hire special teachers, reckoning masters, for arithmetic instruction. While each

Figure 1.2 A merchant surrounded by the tools and wares of his trade, including a counting table.

country had its own name for the reckoning master (in Italy they were called *maestri d'abbaco;* in France; *maistre d'algorisme;* and in the Germanic territories; *Rechenmeister*), they were well-respected in their communities for their computational skills and experience. Cities passed ordinances on how far apart reckoning masters could hang their shingles. The profession of reckoning master became quite lucrative, and often they banded together to form guilds and associations of their own. Many of the reckoning masters were self-employed, a sort of mathematics consultant-at-large, and accepted students for private tuition or conducted formal group classes in their art (see Figure 1.3).[27] Such endeavors gave rise to

Figure 1.3 A 1535 woodcut print showing a father apprenticing his son to a reckoning master. This picture was used in [*Petrarch's Mirror of Consolation*] to illustrate the duties of a just guardian.

reckoning schools whose numbers rapidly proliferated in the commercial cities and along the trade routes of Europe. In 1338, Florence had six reckoning schools, and by 1613 Nuremberg alone could boast 48 such institutions.[28]

Students of the reckoning schools were usually sons of merchants or civil servants, children of the middle class, and probably were between the ages of twelve and sixteen. While, at this time, there were two basic pre-university schools in which arithmetic was taught, the reckoning school and the Latin grammar school, the main purpose of the latter, as its name implies, was to teach Latin—the language of the "educated man." For the students of the Latin schools, mathematics, that is, arithmetic, was merely an extension of the classics; computation, even when taught with use of Hindu-Arabic numerals, was rather theoretical and devoid of societal applications.

So it was to the reckoning schools of Italy that young merchants travelled from all over Europe to learn the *Welsche Praktik,* the mathematical arts, of their trade. The seriousness of this task is emphasized by the advice of a German father to his son studying in Venice, to "rise early, go to church regularly and to pay attention to his arithmetic teacher."[29] Many students of these early reckoning schools, for example, Tartaglia (c. 1499–1557), went on to become noted reckoning masters and mathematicians in their own right.

The Italian Reckoning School

An appreciation of the institution of a reckoning school is rather central to obtaining an understanding of the possible use and impact of the *Treviso Arithmetic.* With a breakdown of many social institutions following the fall of the Roman Empire, formal educational provisions for young people diminished; however, they did not collapse completely. In Italy, existing models for primary education were modified to meet the demands of the prevailing Christian milieu.[30] A system of basic grammar schools emerged that supplied instruction in rhetoric: reading and writing, and simple arithmetic employing Roman numerals. One principal text for such schools consisted of the writings of Donatus, a fourth-century Roman grammarian. In most urban locales, a variety of schools existed: either public—sponsored by the municipal council—or private, or church-related. They supplied a functional literacy in both the vernacular language and Latin. Children entered this system of schooling between the ages of five and seven and completed the course of study by age eleven or twelve. If they then aspired to a profession such as law, medicine, or theology, the services of a higher grammar school were available, where a rigorous study of the Roman masters took place and their literacy and rhetorical skills were sharpened. It was through such a system of schooling, and the existence in Italy of extensive opportunities for the laity to enter a profession that, by the late Middle Ages, the country could boast of having a more highly-educated populace than any of its contemporaries north of the Alps.[31] By the dawn of the twelfth century this system of education was vigorous enough to give rise to the establishment of universities.

Traditional methods of education for entrance into a trade relied mainly on the institution of apprenticeship, under which a youth learned the required skills and knowledge of the desired profession either from family members or a respected master. In the former case, one did not consider such interaction primarily a means of education *per se,* as the child was merely entering the family business as expected. Entrance into the family business was pragmatically motivated, as the child's free labor contributed to the economic growth of the business. In the instance of apprenticeship to a master outside of the family, economic benefits were also reaped by the master in the form of cheap labor and, if the master's trade was highly desirable or his reputation noteworthy, a stipend would also be paid by the apprentice's parents. This latter situation denotes the existence of a truer pupil–teacher relationship. With the urbanization of European society, opportunities in commerce and manufacturing multiplied rapidly and the training of apprentices took on both a new urgency and complexity. Expanding businesses, particularly those involving crafts, needed large numbers of personnel, more than could be accounted for by in-family training, and the knowledge required to function efficiently in many positions became more sophisticated. In some professions, guilds assisted in and supervised the training of apprentices but, in others, the learning of special knowledge still required the searching out of an individual teacher and the securing of his services. So it was with the learning of mathematics.

As the activities of the merchant profession moved from the limited scope of the itinerant peddler to the entrepreneurship of the international commercial house, preparation for entry into the business world became more prolonged and rigorous.[32] A merchant had to be literate, if not in several languages, at least in his own; therefore, boys aspiring to the merchant profession attended the basic grammar schools. Then, upon securing a fundamental literacy and numeracy they advanced onward at ages 11 to 12 to a special secondary school to study commercial arithmetic, or as the Italians phrased it, "moved on to study the abacus."[33] The existence of these *scuola d'abbaco,* schools of the abacus, was a testimony to the importance of mathematics and computational proficiency in the business world of the time. Quite simply, merchant apprentices had to bring with them a special knowledge of mathematics upon entering their apprenticeship.[34] Then, while practicing the merchant trade at the novice state, they perfected and expanded their mathematical techniques.

The appearance of these reckoning schools began in the commercial cities of Italy prior to the fourteenth century and spread northward along the trade routes into Europe. Each school functioned around the knowledge and experiences of a reckoning master who, alone or together with assistants, instructed youths in the content and techniques of commercial arithmetic. In Italy, these institutions of learning were popularly called *botteghe,* which is a word used to denote a shop-house of the period, that is, a structure that housed a business on the ground floor and provided living quarters above. Maestri d'abbaco probably rented such shops and conducted classes of instruction in them. A few surviving documents sketchily supply us with information concerning the organization and curriculum of *botteghe.*

Codex 2186 of the Riccardeana Library in Florence provides some insights into the content taught and the didactics employed in a *bottegha.*[35] In this manuscript, a Pisan, Christofano di Gherando di Dino, describes the syllabus followed in his reckoning school in the year 1442. Students were first taught the writing of the numerals 9, 8, 7, 6, 5, 4, 3, 2, 1. They then learned how to record numbers on their fingers, with fingers of the left hand representing units and those of the right powers of ten. With this preliminary knowledge mastered, the boys moved into a consideration of multiplication. The study of multiplication consisted of the memorization of many tables, or *librettine,* of multiplication facts covering the products up to 99 × 99. Included in this task was a consideration of special products the master thought important; for example, when the multiplier was a prime number less than 50 or a multiple of 10. Next, a study of division was undertaken in a similar manner. Work with practical problems involving discount and interest were given to the students in quantities the master thought would benefit his charges. Gherando di Dino also advocated homework to be assigned each evening, according to the student's ability, and brought in the following morning. Some instructional consideration was given to the study of fractions and the computation of area as associated with land survey. While the description of the actual curriculum is brief, one receives the impression that mathematical instruction for the Pisan school relied heavily on rote memorization and extensive drill work.

A clearer understanding of the mechanics of *bottegha* instruction is supplied by the accounts of a Florentine school run by a master, Francesco di Leonardo Ghaligai, in the year 1519.[36] Ghaligai's school accommodated boys between the ages of eleven and fourteen. Its course of instruction lasted about two years. The curriculum consisted of seven *muta,* or parts, which were to be taken consecutively, after which further instruction by the master, if so desired, could be contracted. Students paid separately for each *muta* completed, the fee being rendered at the end of the instruction. Classes met for six days a week, with the exception of feast days or church holidays. Sessions were held both in the mornings and in the afternoons. As in the Pisan situation, a heavy instructional emphasis was placed on the memorization of facts and the use of algorithmic procedures. Ghaligai's curriculum is briefly outlined as follows:

Muta
1. Multiplication
   a) Memorization of multiplication facts
   b) Practice in the use of algorithms
2. Division by a single digit number
3. Division by a two digit divisor
4. Division by three or more digits
5. Fractions
   a) Basic operations
   b) Use in problem situations
6. The Rule of Three
7. Principles of the Florentine monetary system

It is not clear whether such a school taught the principles of bookkeeping. Perhaps this knowledge was saved for "on-the-job" training, where close supervision could be assured.

After completing his *bottegha* education, a youth began an apprenticeship in which his mathematical knowledge would be greatly broadened and sharpened by the demands of the workplace. Some alumni of *botteghe* went on to become *maestri d'abbaco* of renown in their own right, founded *botteghe,* and even wrote texts on the subject of arithmetic. Thus, through the institution of the reckoning school, a nucleus of practical mathematicians was formed in late medieval and early Renaissance Europe.

Printing and Early Arithmetic Books

The *Treviso Arithmetic* was printed in 1478, making it an incunabula, literally a cradle book, an example of European printing dating before 1501. The typographical compilation of mathematics books with numerical diagrams provided a severe test of early printers' ingenuity. *Treviso* is the earliest known printed mathematics book in the West, and one of the first printed European textbooks dealing with a science.[37] At times, the more comprehensive and influential *Summa de arithmetica geometria proportioni et proportionalita*[38] of Luca Pacioli, printed in Venice in 1494, has been credited with this claim; however, it is clearly predated by the Treviso work and others. The first printed, dated, arithmetic in Germany, appeared in 1482; in France and Spain, 1512; in Portugal, 1519, and in England, 1537.[39] All of these arithmetics were of a commercial type and many were written by reckoning masters. The fact that the classicists demanded a printed edition of Euclid,[40] and that this priority was superseded by the appearance of common arithmetics, tells much about the real mathematics climate of this time. Practical necessity was the motivating force in this printing decision, as indicated by the words of Gaspar Nicolas, the author of the 1519 Portugese book, in his dedication:

> I am printing this arithmetic because it is a thing so necessary in Portugal for transactions with the merchants of India, Persia, Arabia, Ethiopia, and other places discovered by us.[41]

The first printed edition of Euclid appears from the Venetian press of Erhard Ratdolt in 1482. Perhaps the typographical problems inherent in setting type for geometrical figures were responsible for the delay, but more likely it was due to the economic and intellectual demands of the marketplace.[42]

Another noteworthy fact about the *Treviso Arithmetic* is that, as the first printed arithmetic, it appears in the Venetian Republic—not in Germany, the home of printing. This observation can also be extended to Ratdolt's edition of Euclid, further attesting to the Venetians' involvement with mathematics.

While the printing of books employing movable type originated in Mainz with Gutenberg and Fust in about 1450, it rapidly spread to other cities—Cologne, Strasburg, and Venice. Indeed, the Italian cities, including Venice, quickly attracted the practitioners of Gutenberg's new technique. Learning was valued in Italy, the seat of the Renaissance; there was an abundance of manuscript collections to be set to type, and a ready supply of paper and an availability of rich patrons, both ecclesiastic and secular, to purchase printed editions of cherished classics. The first printed Italian book, *Decor Puellarum,* appeared in 1461 and was devoted to a discussion on the upbringing of young girls.[43] Its printing was the work of Nicolas Jensen, a Frenchman who learned his trade in Mainz, ostensibly to serve the French court, but was beckoned by the opportunities of the South and emigrated to Venice to ply his new art. Jensen's efforts marked the beginning of the Venetian printing industry which, by the end of the fifteenth century, encompassed 268 printing establishments and saw two million volumes produced.[44] Venetian printers were not only prolific, but also innovative—books produced in Venice soon exhibited new and novel features, for example, page numbers. The city claimed another industry for its commercial conglomerate.

Treviso, itself, also attracted printers, possibly for its available paper supply. It is known that, before 1500, there were at least six printing establishments in the city: those of Bernardo Colonia (1477), Johannes Rubeus (1485), Bartholomaeus Confalonerius (1483), Joannes de Hassia (1476), Gerardus de Lisa, and one Michael Manzolo or Manazolus.[45] It is from the last of these that the *Treviso Arithmetic* appeared in 1478. Manzolo operated a printing establishment in Treviso from 1476 to 1481, after which he moved his business to Venice. Its author remains anonymous, not an unusual feature for either early arithmetics or works produced in the Venetian Republic. From the opening comments of the text, we do know that the author is a *maestro d'abbaco* and a well-respected member of the community. He styles his work a *practica,* a name quite common in speaking of Italian commercial arithmetics of the early Renaissance.

Up until this time, European arithmetics were of four basic types: theoretical tracts, algorisms, abacus arithmetics, and computi. The theoretical works which provided a basis for scholastic mathematical contemplation were modeled after the writings of Boethius (c.475–524), whose sixth-century *De institutione arithmetica libri duo* was based on translations of Nicomachus of Gerasa (c.100), a neo-Pythagorean mathematician, and Euclid.[46] They presented concepts of Pythagorean number mysticism, such as figurative numbers and the theory of proportions. Boethius' work, in turn, served as a basis for the arithmetical writing of such scholars as Senator Cassiodorus (490–585), Isidorus (c.610), and Jordanus Nemorarius (c.1225). The last known of the Boethian arithmetics was published in Paris in 1521. Boethian arithmetics were the primary source of all arithmetic taught in the Latin schools and universities, at least to the end of the twelfth century, and were highly respected even after the Hindu-Arabic system of numeration and computation was introduced into Europe.

The words "algorism" and "algorithm" owe their etymological origins to the name of the Muslim scholar and author Abu Jafar Muhammed ibn Musa al-Khwarizmi (Muhammed, the father of Jafar and the son of Musa, the Khwarizmian, c. 825). Among his writings, al-Khwarizmi produced an arithmetic on the Hindu numerals and their computational schemes.[47] This arithmetic found its way to Spain where, in the twelfth century, it was translated into Latin by an Englishman, Robert of Chester, and bore the title *Algoritmi de numero indorum.* The manuscript begins with these words:

> Dixit Algoritmi: Laudes deo rectori nostro atque defensori dicamus dignas
> [Algoritmi has spoken: praise be to God, our Lord and our Defender][48]

Thus the Latinized name of al-Khwarizmi used in this typical Islamic salutation became associated with the new numerals and methods of computation. The *Salem Codex,* a German monastery manuscript dating from about the year 1200, further shows how this nominal transition was taking place. It opens:

> Here begins the book of Algorithmus. All wisdom and knowledge comes from God our Lord. . . .[49]

The *Carmen de algorismo,* or Song of the Algorismus, written in verse by the French monk Alexandre de Villedieu (c. 1225), firmly establishes the name of the new numerical techniques:

> Here begins the algorismus.
> This new art is called the algorismus, in which out of these twice fine figures
> 0987654321
> of the Indians we derive such benefit. . . .[50]

While de Villedieu's mathematics is correct, he confused the origins of the word *algorismus,* and erroneously attributed it to the name of an Indian king, Algor. Thus, the technical term *algorism* became applied to any small treatise which explained how to use Hindu-Arabic numerals.[51] Other early algorisms of note are Johannes de Sacrobosco's (c. 1230) *Algorismus vulgaris,* and the anonymous *The Crafte of Nombrynge* (c. 1300). *The Crafte of Nombrynge* was the first algorism to appear in England, and was an interpretation of the *Carmen de algorismo.* It also confused the origin of the word *algorism:*

> This book is called the book of Algorism, and this book treats the Craft of Numbering, which Craft is called also Algorism. There was a King of India, the name of whom was Algor, and he made this his craft, and, after his name, he called it Algorism.[52]

During the twelfth through the fifteenth centuries, algorisms appeared in great numbers and in a diversity of languages other than Latin; including Hebrew, Italian, French, and Icelandic.[53] An algorism whose problem situations dealt with the affairs of business and commerce was also called a *practica.*[54]

Like the practicae, some abacus arithmetics also concerned commercial reckoning, but their computational methods were based on the use of Roman numerals and a concrete computational scheme involving the manipulation of counters (*calculi* in Latin, or *jetons* in French) on a table upon which place value columns or rows had been inscribed. This computing table, usually found in the commercial houses of the Middle Ages, evolved into the popular "counter" over which modern-day business is transacted; however, the table was really a form of abacus, and the techniques of using it comprised abacus arithmetic.[55] The word *abacus* is Latin, and is derived from the Greek term αβαξ, *abax* (slab) which, in turn, is related to the Hebrew word for dust, *abhaq*. In antiquity, the computing table was a slab covered with fine dust on which computation was undertaken and easily erased; thus, a device for computing became known as an abacus. Even when the dust was replaced by counters and ruled lines, the name remained. In the late Middle Ages and early Renaissance, a bitter controversy raged between the advocates of the Hindu-Arabic system—the algorists—and the abacists. The rivalry between these two mathematical elements was noted in the popular thirteenth-century knightly romance, *Jüngere Titurel:*

Figure 1.4 A typical counting table of the time.

And here too comes out
Lot, Prince of Norway,
With I know not how many hundreds;
If Algorismus were still alive
And Abakuc, learned in geometry,
They would have much to do
To find the number of them all[56]

and in many woodcuts of the time (see figure 1.5). Tradition was not easily moved.[57]

Figure 1.5a Imaginative woodcut print illustrating a contest between abacists and algorists. The spirit of arithmetic, in female form, presides over the competition (1508).

Figure 1.5b The title page of Adam Riese's *Rechenbuch* (1529) illustrates the computing controversy of his time.

Examples of abacus arithmetics are *Regulae de numerorum abaci rationibus,* by Gerbert of Aurillac (c.1003); Adelard of Bath's *Regule abaci* (c.1120); Ralph of Laon's *Liber de abaco;* and *Regunculi super abacum* (c.1200) by Turchillus. By the time of the early Renaissance, abacus arithmetics were losing their popularity in Italy to algorisms; however, they were still widely used in northern Europe, where abacus reckoning persisted.[58] As late as 1592, the abacus computing table was still used in Fugger merchant houses.[59]

Computi were brief treatises devoted to computation of the ecclesiastical calendar and were more descriptively known as *Computus paschalis* or *Computus ecclesiasticus*. They gave directions on mathematically determining the dates of Easter and other movable church feasts.[60] Early examples of such works were *Paschalis sive de indicationibus cyclis solis et lunae* (562), attributed to Cassiodorus, and the Venerable Bede's (c.673–735) *De temporum ratione*. Present-day almanacs are descendants of computi.

The *Treviso Arithmetic* is an algorism, a practica, intended for self study and relevant to the commercial and reckoning needs of Treviso and Venetian trade. It is written in the Venetian dialect, a feature attesting to its egalitarian mission of communicating knowledge to a broad audience. Throughout the Middle Ages and well into the Renaissance, Latin was the language of scholarly instruction and education, thus limiting great portions of knowledge to a select population. Vernacular texts such as the *Treviso* eliminated a monopoly on knowledge and gave great impetus to the rise of a middle class. From the thirteenth century onward, thousands of students at the universities had a Latin exposure to the new arithmetic, but it was not until the acceleration of commercial activities and the advent of printing that this knowedge was disseminated to the "common man." It has been estimated that between the origin of European printing and the end of the fifteenth century, thirty practical arithmetics were printed, of which more than one-half were written in Latin, seven in Italian, four in German, and one in

French.[61] The numbers of vernacular texts supply a relative index of the mercantile activity taking place in their respective countries at that time. During the same period, about 26 theoretical Boethian-style Latin arithmetics were also produced.[62] Mathematics was moving from the realm of scholastic speculation to the applications of manufacture and the marketplace.

Some Features of the Treviso Arithmetic

The *Treviso Arithmetic* was not written for a large audience. Its problem situations, while specific to the times, apparently were not comprehensive enough to make it a popular practica for commercial arithmetic. Particularly, its consideration of monetary exchange might be considered meager.[63] There appears to have been only one edition of the work, and extant copies are extremely rare. The only early bibliographers known to have come into personal contact with a copy of the book were Domenico Maria Federici and Baldassarre Boncompagni. Federici describes it in his *Memorie Trevigiani* (1805); however, Federici's copy has since disappeared. Boncompagni catalogued it in *Atti dell'Accademia Pontifica de'Nuovi Lincei* (1862–63), and included facsimiles of some of its pages. The copy of the *Treviso Arithmetic* used for this study is housed in the Plimptom Collection, Rare Books and Manuscript Library, of Columbia University.[64] The volume found its way to this collection via a circuitous route. Maffeo Pinelli (d. 1785), an Italian bibliophile, is the first known owner.[65] After his death, his library was purchased by a London book dealer and sold at auction on February 6, 1790.[66] The book was obtained for three shillings by a Mr. Wodhull.[67] About 100 years later, the arithmetic appeared in the library of Brayton Ives, a New York lawyer. When Ives' collection of books was eventually sold at auction, George A. Plimpton, a New York publisher, acquired the *Treviso Arithmetic* and made it an acquisition to his extensive collection of early scientific tracts. Plimpton donated his library to Columbia University in 1936.

The book is a quarto work; i.e., each printed sheet forms four pages. There are 123 actual pages of text, with 32 lines of print to a page. The pages are unnumbered, untrimmed, and have wide margins. Some of the margins contain penned-in notes of past users. The size of the book is 14.5 cm by 20.6 cm, a small tome by the standards of early Renaissance printers, indicating it was designed for popular use. Numerous mathematical diagrams are scattered throughout the text illustrating algorithmic procedures; however, the book contains no pictorial illustrations.

✱ 2
A Translation of the Treviso Arithmetic

Prologue

The material in the section below is a translation of the *Treviso Arithmetic,* 1478, in its entirety. David Eugene Smith completed the translation in 1907 and revised it in 1911. Although he used some of the translated material for a 1924 article in *Isis* and made references to the Treviso manuscript in several of his works,[1] the complete translation has not been previously published.

Among the many competencies credited to Smith was his ability as a linguist and translator,[2] but he was primarily a mathematics educator and his rendering of the *Treviso Arithmetic* is a free translation offered as a service to students of the history of mathematics. As he explains:

> Such a work, epoch-making as it was in the history of education, deserves the attention of all who are interested in the

Jncomminciа vna practica molto bona et vtile a ciascheduno chi vuole vxare larte dela merchadantia.chiamata vulgarmente larte de labbacho.

Pregato piu e piu volte da alchuni zouani a mi molto dilectissimi : li quali pretendeuano a douer volere fare la merchadantia:che per loro amore me piacesse affadigarme vno puocho:de dargli in scritto qualche fundaméto cerca larte de arismetrica:chiamata vulgarmente labbacho.Unde io constretto per amor di loro: et etiádio ad vtilita di tuti chi pretendano a quella:segondo la picola intelligentia del inzegno mio:ho deliberato se non in tuto:in parte tanté satisfare a loro.acio che loro virtuosi desiderii vtile frutto receuere posseano.Jn nome di dio adoncha : toglio per principio mio el ditto de algorismo cosi dicédo.

t Ute quelle cose:che da la prima origine hano habuto produciméto:per ratone de numero sono sta formade.E cosi come sono:hano da fir cognoscude.Pero ne la cognitione de tute le cose:questa practica e necessaria . E per intrar nel pposito mio:primo sapi lectore:che quanto fa al proposito nostro:Numero e vna moltitudine congregata ouero insembrada da molte vnitade.et al meno da do vnitade.come e.2.el quale e lo primo e menore numero:che se truoua.La vnitade e quella cosa : da la quale ogni cosa si ditta vna.Segódario sapi:che se truoua numeri de tre maniere.El primo se chiama numero simplice.lal tro numero articulo . El terzo se chiama numero

Libro de Marco Belerami

Figure 2.1 The first page of the *Treviso Arithmetic.* At the bottom of the text, an early owner proudly claimed possession by affixing his name, Marco Belerami. A problem involving monetary conversions has been written in the right-hand margin.

> evolution of teaching. It is, however, so rare as to be inaccessible to most students, and for this reason it has seemed to the translator a matter of duty to make it available for study. Having known the difficulties of reaching the original sources in the history of mathematics it is his purpose to do what he can to make it possible for others to save much time that he has had to lose in this effort. It is for this reason that he offers to those who do not care to study the Venetian dialect of the early Renaissance and who have not access to the book itself, a free translation of the work. It was the purpose, originally, to make a translation that should preserve many of the curious archaic forms of the original; but on reflection, this was felt to be unwise. The effort has, therefore, been made to give the meaning of the text, without attempting a verbatim translation. Not being a contribution to the domain of philology, but rather to the history of mathematics, the translation seems best to serve its purpose if undertaken in this spirit.[3]

The translation is mathematically and historically correct and its wording captures the spirit and purpose of the *Treviso*'s author. Evidence of human error and limitations have been preserved in the translation—several mistakes are evident; accordingly, corrections are indicated by the use of square brackets. Typographical anomalies are not infrequent; for example, a lack of type fonts for the numeral "1" forced the use of "i" in its place. Such printing errors will be retained in some of the text. Certain editorial liberties have been taken for the benefit of a modern reading audience, as when in a problem the word *lira* may be used both as a monetary measure and the measure of a commodity weight.[4] The word *lira* will be retained for money but the English word "pound" will be used to denote weight. Pagination has been introduced into the translation for the convenience of referencing; the sequential ordering of the folio pages are given by appropriate numbers and letters; i.e., f3v, f3r indicates the third folio left side, verso, and the right side, recto, respectively. In order to preserve the continuity of the presentation, the use of extensive annotation within this chapter has been avoided. Corresponding analysis and discussion is undertaken in subsequent chapters. Specifically, the material of f1r to f14r is considered in Chapter 3; f14r to f22v, Chapter 4; f22v to f43r, Chapter 5; and f44v to f62r, Chapter 6.

The Text

F. 1, r.

Here beginneth a Practica, very helpful to all who have to do with that commercial art commonly known as the abacus.

I have often been asked by certain youths in whom I have much interest, and who look forward to mercantile pursuits, to put into writing the fundamental principles of arithmetic, commonly called the abacus. Therefore, being impelled by my affection for them, and by the value of the subject, I have to the best of my small ability undertaken to satisfy them in some slight degree, to the end that their laudable desires may bear useful fruit.

Therefore in the name of God I take for my subject this work in algorism, and proceed as follows.

All things which have existed since the beginning of time have owed their origin to number. Furthermore, such as now exist are subject to its laws, and therefore in all domains of knowledge this Practica is necessary. To enter into the subject, the reader must first know the basis of our science. Number is a multitude brought together or assembled from several units, and always from two at least, as in the case of 2, which is the first and the smallest number. Unity is that by virtue of which anything is said to be one. Furthermore be it known that there are three kinds of number, of which the first is called a simple number, the second an article, and the third a composite or mixed number. A simple number is one that contains no tens, and it is represented by a single figure, like i, 2, 3, etc. An article is a number that is exactly divisible by ten, like i0, 20, 30, and similar numbers. A mixed number is one that exceeds ten but that cannot be divided by ten without a remainder, such as i1, i2, i3, etc. Furthermore be it known that there are five fundamental operations which must be understood in the Practica, namely numeration, addition, subtraction, multiplication, and division. Of these we shall first treat of numeration, and then of the others in order.

F. 1, v.

Numeration is the representation of numbers by figures. This is done by means of ten letters or figures, as here shown. Of these the first figure, i, is not called a number but the source of number. The

.i.
.2.
.3.

tenth figure, o, is called cipher or 'nulla', .4.
i. e. the figure of nothing, since by itself .5.
it has no value, although when joined .6.
with others it increases their value. .7.
Furthermore you should note that when .8.
you find a figure by itself its value .9.
cannot exceed nine, i. e., 9; and from .0.
that figure on, if you wish to express a number you must use at least two figures thus: ten is expressed by io, eleven by ii, and so on. And this can be understood from the following figures. To understand

F. 2, r.

the figures it is necessary to have well in mind the following table:

i	time	i	makes	i
i	time	2	makes	2
i	time	3	makes	3
i	time	4	makes	4
i	time	5	makes	5
i	time	6	makes	6
i	time	7	makes	7
i	time	8	makes	8
i	time	9	makes	9
i	time	0	makes	0

i	time	io	makes	io
2	times	io	makes	20
3	times	io	makes	30
4	times	io	makes	40
5	times	io	makes	50
6	times	io	makes	60
7	times	io	makes	70
8	times	io	makes	80
9	times	io	makes	90
0	times	io	makes	0

i	time	i00	makes	i00
2	times	i00	makes	200

F. 2, v.

3	times	ioo	makes	300
4	times	ioo	makes	400
5	times	ioo	makes	500
6	times	ioo	makes	600
7	times	ioo	makes	700
8	times	ioo	makes	800
9	times	ioo	makes	900
0	times	ioo	makes	0

Thousands of millions	Hundreds of millions	Tens of millions	Millions	Hundreds of thousands	Tens of thousands	Thousands	Hundreds	Tens	Units
									i
									2
									3
									4
									5
									6
									7
									8
									9
								i	0
							i	2	0
						i	2	3	0
					i	2	3	4	0
				i	2	3	4	5	0
			i	2	3	4	5	6	0
		i	2	3	4	5	6	7	0
	i	2	3	4	5	6	7	8	0
i	2	3	4	5	6	7	8	9	0
2	3	4	5	6	7	8	9	0	0
3	4	5	6	7	8	9	0	0	0
4	5	6	7	8	9	0	0	0	0
5	6	7	8	9	0	0	0	0	0
6	7	8	9	0	0	0	0	0	0
7	8	9	0	0	0	0	0	0	0
8	9	0	0	0	0	0	0	0	0
9	0	0	0	0	0	0	0	0	0

F. 3, r.

And to understand the preceding table it is necessary to observe that the words written at the top give the names of the places occupied by the figures beneath. For example, below "units" are the figures designating units, below "tens" are the tens, below "hundreds" are the hundreds, and so on. Hence if we take each figure by its own name, and multiply this by its place value, we shall have its true value. For instance, if we multiply 1, which is beneath the word "units," by its place, that is by units, we shall have "1 time 1 gives 1," meaning that we have one unit. Again, if we take the 2 which is found in the same column, and multiply by its place, we shall have "1 time 2 gives 2," meaning that we have two units. Furthermore, multiplying the 3 in the same column by its place, we have "1 time 3 gives 3," signifying three units, and so on for the other figures found in this column. Passing to the tens we see that 9 is found in the column directly beneath this word, and 9 times 10 makes 90 shows that this 9 represents ninety. In the same way the 8 represents 8 times 10 or 80, thus the 8 stands for eighty. Similarly with 7, the 7 times 10 making 70 shows that this 7 stands for seventy. This rule applies to the various other figures, each of which is to be multiplied by its place value.

And this suffices for a statement concerning the "act" of numeration.

F. 3, v.

Having now considered the first operation, that is numeration, let us proceed to the other four, which are addition, subtraction, multiplication, and division. To differentiate between these operations it is well to note that each has a characteristic word, as follows:

Addition has the word and.
Subtraction has the word from.
Multiplication has the word times.
Division has the word in.

It should also be noticed that in taking two numbers, since at least two are necessary in each operation, there may be determined by these numbers any one of the above-named operations. Furthermore, each operation gives rise to a different number, with the exception that 2 times 2 gives the same result as 2 and 2, since each is 4. Taking, then, 3 and 9 we have:

Addition:	*3*	*and*	*9*	*makes*	*12*
Subtraction:	*3*	*from*	*9*	*leaves*	*7* [6]
Multiplication:	*3*	*times*	*9*	*makes*	*27*
Division:	*3*	*in*	*9*	*gives*	*3*

We thus see how the different operations with their distinctive words lead to different results.

In order to understand the second operation, addition, it is necessary to know that this is the union of several numbers, at least of two, in a single one, to the end that we may know the sum arising from this increase. It is also to be understood that in the operation of adding, two numbers at least are necessary, namely the number to which we add the other, which should be the larger, and the number which is to be added, which should be the smaller. Thus we always add the smaller number to the larger, a more convenient plan than to follow the contrary order, although the latter is possible, the result being the same in either case. For example, if we add 2 to 8 the sum is 10, and the same result is obtained by adding 8 to 2. Therefore, if we wish to add

F. 4, r.

one number to another we write the larger one above and the smaller one below, placing the figures in convenient order, i. e., the units under units, tens under tens, hundreds under hundreds, etc. We always begin to add with the lowest order, which is of least value. Therefore, if we wish to add 38 to 59 we write the numbers thus:

	59
	38
Sum	97

We then say, "8 and 9 make 17," writing 7 in the column which was added, and carrying the 1 (for when there are two figures in one place we always write the one of lower order and carry the other to the next higher place). This 1 we now add to 3, making 4, and this to 5, making 9, which is written in the column from which it is derived. The two together make 97.

The proof of this work consists in subtracting either addend from the sum, the remainder being the other. Since subtraction proves addition, and addition proves subtraction, I leave the method

F. 4, v.

of proof until the latter topic is studied, when the proof of each operation by the other will be understood.

Besides this proof there is another. If you wish to check the sum by casting out nines, add the units, paying no attention to 9 or 0, but always considering each as nothing. And whenever the sum exceeds 9, subtract 9, and consider the remainder as the sum. Then the number arising from the sum will equal the sum of the numbers arising from the addends.

For example, suppose you wish to prove the following sum:

.59.
.38.
Sum .97. | 7

The excess of nines in 59 is 5; 5 and 3 are 8; 8 and 8 are 16; subtract 9 and 7 remains. Write this after the sum, separated by a bar. The excess of nines in 97 is 7, and the excess of nines in 7 equals 7, since neither contains 9. In this way it is possible to prove the results of any addition of abstract numbers or of those having no reference to money, measure, or weight. I shall show you another plan of proof, according to the nature of the case. If you have to add 816 and 1916, arrange the numbers as follows:

F. 5, r.

1916
816
Sum 2732

Since the sum of 6 and 6 is 12, write the 2 and carry the 1. Then add this 1 to that which follows to the left, saying: "1 and 1 are 2, and the other 1 makes 3." Write this 3 in the proper place, and add 8 and 9. The sum of 8 and 9 is 17, the 7 being written and the 1 carried to add to the other 1, making 2, which is written in the proper place, the sum now being complete. If you wish to prove by 9, arrange the work thus:

1916
816
The sum 2732 | 5

You may now effect the proof by beginning with the upper number, saying: "1 and 1 are 2, and 6 are 8, and 8 are 16. Subtract 9, and 7 remains. The 7 and 1 are 8, and 6 are 14. Subtract 9 and 5 remains," which should be written after the sum, separated by a bar. Look now for the excess of nines in the sum: 2 and 7 are 9, the excess being 0; 3 and 2 are 5, so that the result is correct.

If you have to add 2732 and 45318, the work will appear as follows:

	45318
	2732
The sum	48050

We begin with 2 and 8, which make 10, writing the 0 and carrying 1; 1 and 3 are 4, and 1 are 5; we write this in its proper place; then 7 and 3 are 10, and we write the 0 and carry the 1, adding it to 2, making 3, and this to 5, making 8, which is written in its due place. The 4 in the next column is written as usual, and the sum is 48050.

F. 5, v.

If you wish to prove this work by casting out 9s, arrange it as follows:

	45318
	2732
The sum	48050 | 8

Beginning with the larger number, 4 and 5 are 9, which we may consider 0; 3 and 1 are 4, and 8 are 12; and 12 less 9 leaves 3; 3 and 2 are 5, and 7 are 12, and 3 are 15, and 15 less 9 leaves 6, and 6 and 2 are 8. Write this 8 after the sum, separated by a bar. It is now necessary to see that the excess in the sum is also 8; hence we say: 4 and 8 are 12, and 5 are 17, and 17 less 9 is 8, so that the work is correct. In the same way we can prove any other addition, however many numbers there are. Now that we have learned how to add abstract numbers we shall proceed to the addition of concrete numbers.

We shall consider cases involving lire, soldi, and pizoli, but first the case of lire alone, without soldi and pizoli. Then we shall take that of lire and soldi without pizoli, and then the one involving all three. In order to appreciate better the adding of soldi and pizoli it should be understood that you will never find, for any column, a sum with more than two figures, of which the smaller is called units and the other the tens. Then if we wish to add 569 lire and 392 lire, we arrange the work as follows:

	lire	569
	lire	392
The sum	lire	961 | 7

We then begin to add: 2 and 9 making 11, of which we write 1 and carry 1, which we add to 9, making 10, and this to 6 making 16. We write 6 thus found, and carry the 1. This we add to 3, making 4, and this to 5, making 9, and this gives us the sum, which amounts to 961 lire.

F. 6, r.

If you wish to prove this result, cast out the 9s of both numbers added, saying: "5 and 6 are 11; subtract 9, leaving 2; then 2 and 3 are 5, and 2 are 7," which is the excess, and which is written after the sum, separated by a bar. Now find if the excess in the sum is 7, saying: "6 and 1 are 7," which shows that the work is right.

If you have to add 916 lire, 14 soldi, and 1945 lire, 15 soldi, arrange the work as follows:

	lire	1945	s. 15
	lire	916	s. 14
Sum	lire	2862	s. 9 | 0

Add first the number of soldi, saying: "4 and 5 are 9." Write this in its proper place. Then take the tens of the soldi, noting the nature of the case, whether the number is even or odd. If the number is odd, you should write 1 in the tens' place, taking half of the rest and calling them lire. If the number of tens is even, you should at once take the half, writing it as lire, carrying it to the first figure of the lire and adding it in. Now add the tens, saying: "1 and 1 are 2; half of 2 is 1, which is 1 lire and which is added to the lire; 1 and 6 makes 7 and 5 makes 12," of which we write the 2 and carry the 1. We now add this 1 to the other, saying: "1 and 1 are 2, and 4 are 6," which we write in its proper place. Then 9 and 9 are 18, of which we write the 8 and carry the 1, adding it to the other 1, giving 1 and 1. This 2 we write in its place, and the sum amounts to 2862 lire, 9 soldi.

If you wish to prove this sum, take the excess

F. 6, v.

of 9s in the lire, and multiply this by 20, or by 2; take the excess of 9s in this product, and add it to the excess in the sum of the soldi. In this case we begin by saying: "1 and 4 are 5, and 5 are 10; cancel the 0 and 1 remains; 1 and 1 are 2, and 6 are 8," which is the excess in the lire; multiplying this 8 by the excess in 20, i. e., by

2, we have 2 times 8 are 16; subtracting 9, 7 remains. We now add this 7 to the excess in the soldi, 7 and 1 making 8, and 5 making 13. Subtracting 9, 4 remains; 4 and 1 are 5, and 4 are 9, in which the excess is 0, and this we place after the sum, separated by a bar. We now find if the excess in the sum itself is 0, saying: "2 and 8 are 10; cancel 0 and 1 remains; 1 and 6 are 7, and 2 are 9, with 0 excess; multiply 0 by the excess in 20, and we have 2 times 0 are 0; but the excess in the soldi, 9, is 0," and this shows that the work is right.

If you have to add 892 lire, 15 soldi, and 7 pizoli to 9562 lire, 19 soldi, and 11 pizoli, arrange the work as follows:

	lire	9562	s. 19	p. 11	
	lire	892	s. 15	p. 7	
Sum	lire	10455	s. 15	p. 6	6

Add first all of the pizoli into one sum, saying: "7 and 1 are 8, which with the 1 ten makes 18"; and since there are 12 pizoli in a soldo, we must see how many soldi there are in our sum. We therefore note that 12 is found in 18 once, so that there is 1 soldo, with 6 pizoli over, which 6 we write under the number of pizoli. We then carry this 1 to the number of soldi, and we have: 1 and 5 are 6, and 9 are 15. We write the 5 and carry the 1, adding it to the tens of the soldi, thus: 1 and 1 are 2, and 1 are 3, which 3 is an odd number. We therefore write 1 under the tens and take half of the other 2, which is 1, and call it one lira. This 1 lira we carry to the number of lire, thus: 1 and 2 are 3, and 2 are 5, which 5 we write in its

F. 7, r.

place. Then 9 and 6 are 15, and we write the 5 and carry the 1. Then 1 and 8 are 9, and 5 are 14, of which we write the 4 and carry the 1. This 1 we add to the 9, obtaining 10, and write the 0 under the 9 and the 1 to the left. The sum therefore amounts to 10455 lire, 15s., 6p.

If you wish to prove this, take the excess of 9s in the lire, thus 5 and 6 are 11, and 2 are 13; subtract 9, and 4 remains; 4 and 8 are 12, and 2 are 14; subtract 9, and 5 remains; multiply this 5 by the excess in 20, i.e., by 2, and 2 times 5 are 10, with an excess of 1; then add this 1 to the excess in the soldi, thus: 1 and 1 are 2, and 1 are 3, and 5 are 8; then multiply this 8 by the excess in 12; i.e., by 3, and 3 times 8 are 24; the excess in this 24 is 6, since 2 and 4 are 6. Then add this 6 to the excess in the pizoli, thus: 6 and 1 are 7, and 1 are 8, and 7 are 15; subtract 9 and 6 remains. This is the excess in the numbers added, and this is written to the right of the sum, separated by a bar. We now proceed to find if the excess in the sum is 6, thus: 1 and 4 are 5, and 5 are 10, and 5 are 15; subtract 9 and 6 remains; multiply this 6 by the excess in 20, i.e., by 2,and 2 times 6 are 12; subtract 9 and 3 remains. We add this 3 to the soldi, thus: 3 and 1 are 4, and 5 are 9, with 0 excess. Multiply this 0 by the excess in 12, i.e., by 3, and 3 times 0 are 0. There remains only the excess

F. 7, v.

in the pizoli, which is 6, hence the result is correct.

Let us now take up the method of adding ducats, grossi, and pizoli in gold. In order to understand this operation it is necessary to know that 32 pizoli make a grosso, and 24 grossi make a ducat. Therefore, add the pizoli together to make grossi, dividing the total number by 32, and the remainder, which is pizoli, is written under the pizoli. Then add the grossi derived from the pizoli to the other grossi to make ducats, dividing the total number of the grossi by 24, and the remainder will be grossi, which are written in their proper place. Then add the ducats which are derived from the grossi to the other ducats, and so on to the left.

Suppose you have to add the following:

	ducats	2169	g. 23	p. 31
	ducats	1902	g. 16	p. 23
Sum	*ducats*	4072	g. 16	p. 22 | 6

Commence by adding the pizoli together, saying: "3 and 1 are 4," for the units. Then add the tens, thus: 2 and 3 are 5, which equals 50, and which with the 4 makes 54. Now 32 is contained once in 54, whence this equals 1 grosso and 22 remainder, which equals 22 pizoli, written under pizoli. The grosso thus found is added to the other grossi, thus: 1 and 6 are 7, and 3 are 10, of which we write the 0 in the proper place and carry the 1 of the 10 to the tens' place. Then 1 and 1 are 2, and 2 and 4, which 4 with the 0 makes 40. These 40 grossi divided by 24, which is contained once in 40, gives 1 ducat and a remainder of 16 grossi, which are written under the other grossi.

F. 8, r.

Now add this ducat derived from the grossi to the number of grossi [ducats], thus 1 and 2 are 3, and 9 are 12; write 2 and carry 1; then 1 and 6 are 7, which is written in its proper place. Then add the 9 to the 1, thus: 1 and 9 are 10;

write the 0 and carry 1, adding it to the other 1, making 2, and 2 more make 4. Write 4, and you have for the sum

ducats 4072 *g.* 10 *p.* 22

To prove that the result is correct, take the excess in the ducats, thus: 2 and 1 are 3, and 6 are 9, which as an excess of 0; then 1 and 2 are 3, which is the excess of the ducats. Now multiply this 3 by the excess of 24, or 6, having 3 times 6 are 18, in which 1 and 8 are 9, with 0 excess. Then take the excess of the grossi, thus: 2 and 3 are 5, and 1 are 6, and 6 are 12; subtract 9 and 3 remains; this 3 multiplied by the excess of 32, or 5, gives 3 times 5 or 15; subtracting 9, 6 remains, and this 6 is added to the pizoli. Then 6 and 3 are 9, with 0 excess; continuing with the rest, 1 and 2 are 3, and 3 are 6. Then 6 is the excess in the addends, and this we write after the sum, separated by the bar. We now consider whether the excess in the sum is also 6, thus: 4 and 7 are 11; subtract 9 and 2 remains; 2 and 2 are 4, and this 4 multiplied by the excess in 24, i.e., 6, gives us 4 times 6 or 24; the excess here is found by adding 2 and 4, giving 6. This excess is added to the excess of the grossi, thus: 6 and 1 are 7, and 6 are 13; subtracting 9, 4 remains. Now multiply this 4 by the excess in 32, which is 5, and we have 4 times 5 are 20, of which the excess is 2, which must now be added to the pizoli, thus: 2 and 2 are 4, and 2 are 6, which is the excess, and shows that the work is correct. By the same method it is possible to proceed in similar cases.

F. 8, v.

In order to understand the addition of lire and grossi it is necessary to know it for lire, soldi, grossi, and pizoli. For this purpose it is necessary to know that 32 pizoli make 1 grosso, 12 grossi make 1 soldo, and 20 soldi make 1 lira, and in each lira there are 10 ducats. We therefore add two numbers as follows:

	lire	5612	s. 18	g. 11	p. 19
	lire	2820	s. 5	g. 4	p. 8
Sum	lire	8433	s. 4	g. 3	p. 27 | 3

We begin by adding the pizoli, thus: 8 and 9 are 17, and 10 are 27; with 27 pizoli for the sum, there being no grossi in this amount. Then add the grossi, thus: 4 and 1 are 5, and 10 are 15; 12 is contained in 15 once, which gives 1 soldo; 12 from 15 leaves 3 remainder, or 3 grossi, which is written under the other grossi. The 1 soldo is added to the others, thus: 1 and 5 are 6, and 8 are 14; write the 4 and carry 1, which, being tens of soldi, is added to the rest. Then 1 and 1 are 2, and we take half of 2, which is 1, or 1 lira; for tens of soldi always make lire if we divide by two until we get a remainder of 1, which we then leave under the tens' column. We now add this 1, namely the lira, to the others, thus: 1 and 2 are 3, which 3 we write in its proper place. Then 2 and 1 are 3, which we write in its column. Then 8 and 6 are 14, the 4 being written and the 1 being carried. Then 1 and 2 are 3, and 5 are 8, which 8 is written in its proper place. This, then, is the sum.

Sum	lire	8433	s. 4	g. 3	p. 27

If we wish to prove this result we take the excess of the addends thus: 5 and 6 are 11;

F. 9, r.

subtract 9 and 2 remains; 2 and 1 are 3, and 2 are 5, and 2 are 7, and 8 are 15, and 2 are 17; subtract 9 and 8 remains; which is the excess of the lire. Multipy this 8 by the excess in 20, namely 2, and 2 times 8 are 16; subtract 9 and 7 remains. Then add this 7 to the soldi, thus: 7 and 1 are 8, and 8 are 16; subtract 9 and 7 remains; 7 and 5 are 12; subtract 9 and 3 remains, which is the excess of the soldi. Then multiply this 3 by the excess of 12, that is 3, and 3 times 3 are 9, with 0 excess. Now take the excess of the grossi, thus: 1 and 1 are 2, and 4 are 6, which is the excess of the grossi. Multiply this by the excess of 32, i.e., 5, and 5 times 6 are 30. Cancel the 0 and 3 remains, which 3 we add to the pizoli, thus: 3 and 1 are 4, and 8 are 12; subtract 9 and 3 remains, which is the excess of the addends. This we write after the sum to the right of the bar. Now consider if the excess in the sum itself is also 3, thus: 8 and 4 are 12, and 3 are 15; subtract 9 and 6 remains; 6 and 3 are 9, with 0 excess, which cancels the excess in 20, thus: 2 times 0 is 0. Now take the excess of the soldi, which is 4, and multiply this by the excess of 12, which is 3, thus 3 times 4 is 12, subtract 9 and 3 remains. Add this to the excess of the grossi, thus: 3 and 3 are 6, and this 6 multiplied by the excess of 32, i.e., 5, gives 5 times 6 or 30. Cancel the 0 and 3 remains, which proves the result. In the same way we may perform all similar additions, always arranging the work so that the less denominations shall be reduced to the greater ones, as we have seen in the case where the pizoli reduced to soldi, the soldi to lire, etc. This sufficeth for the operation of addition.

Having now considered the second operation of the Practica of arithmetic, namely the operation of addition, the reader should give attention to the third, namely the operation of subtraction. Therefore I say that the operation of subtraction is nothing else than this: that of two

F. 9, v.

numbers we are to find how much difference there is from the less to the greater, to the end that we may know this difference. For example, take 3 from 9 and there remains 6. It is necessary that there should be two numbers in subtraction, the number from which we subtract and the number which is subtracted from it. The number from which the other is subtracted is written above, and the number which is subtracted below, in convenient order, i.e., units under units and tens under tens, and so on. If we then wish to subtract a number of some order from another we shall find that the number from which we are to subtract is equal to it, or greater, or less. If it is equal, as in the case of 8 and 8, the remainder is 0, which 0 we write underneath in the proper column. If the number from which we subtract is greater, then take away the number of units in the smaller number, writing the remainder below, as in the case of 3 from 9, where the remainder is 6. If, however, the number is less, since we cannot take a greater number from a lesser one, take the complement of the larger number with respect to 10, and add this to the other, but with this condition: that you add one to the next left-hand figure. And be very careful that whenever you take a larger number from a smaller, using the complement, you remember the condition above mentioned. Take, now, an example:

Subtract 348 from 452, arranging the work thus:

	452
	348
The remainder	104

First we have to take a greater number from a less, and

F. 10, r.

then an equal from an equal, and third, a less from a greater. We proceed as follows: we cannot take 8 from 2, but 2 is the complement of 8 with respect to 10, and this we add to the other 2 which is above the 8, thus: 2 and 2 make 4, which we write beneath the 8 for the remainder. There is, however, this condition, that to the figure following the 8 (i.e., 4), we add 1, making it 5. Then 5 from 5, which is an equal, leaves 0, which 0 we write beneath. Then 3 from 4, which is a less from a greater, is 1, which 1 we write under the 3, so that the remainder is 104.

If we wish to prove this result, add the number subtracted to the remainder, and the result will be the number from which we subtracted. We may arrange the work as follows:

452	2
348	6
104	5
452	

Now add, 4 and 8 are 12; write 2 under the 4 and carry 1; then 1 and 4 are 5; write this 5 under the 0; then add 1 and 3, making 4, and write this 4 under the 1, and the work checks. Thus is found that which was promised you, as you can see.

Notwithstanding the proof already given, which for neatness leaves nothing to be desired, I wish you to know that you can prove the above result by casting out 9s. Add the digits of the larger number, thus: 4 and 5 are 9, with 0 excess; then take the 2 for the excess of 452, writing it after the number. Then add the digits of 348, as before: 3 and 4 are 7, and 8 are 15; subtract 9 and 6 remains as the excess, which is to be written after this number. Then take this 6 from 2, which being impossible we proceed as I have explained above. Take the complement of 6 with respect to 9, since 9 is the basis of the proof, and you have 3. Add this 3 to the 2 above and you have 5, which is the principal excess, and which is written after the remainder separated by a bar. Then see if the excess in the remainder is also 5, thus: 1 and 4 are 5. We therefore see that the result is correct. In the same way we can prove any other case of subtraction. I do not repeat this in the other examples, but all of the examples which are given in subtraction I prove by addition, because this is more rapid and also more certain than the proofs by 9s.

F. 10, v.

Now having understood the method of subtracting, let us return to addition and prove this by means of subtraction. Let us begin by considering the first example which we had in the chapter on addition, as follows:

	59
	38
The sum	97

Now take 38 from 97 and the remainder is 59, which shows the sum to be correct; or arrange the work as follows:

	97
	38
The remainder	59

Then take 8 from 7; this is impossible; but 2, the complement of 8 with respect to 10, added to 7, makes 9, which remainder of this order we write under

F. 11, r.

the 8. Then carry 1 to the 3, making 4, and we have 4 from 9 leaving 5, which write beneath the 3, and the result is 59. We therefore see the sum is correct. Note, also, that this operation of subtraction needs no proof, since we have for its proof the addition already made, that of uniting 38 and 59, making 97. Now let us prove once more this first sum, and subtract 59 from 97 and if we get 38 for a remainder the sum will be correct. Arrange the work as follows:

	97
	59
The remainder	38

Take 9 from 7; 9 cannot be subtracted from 7, but 1 is the complement of 9 with respect to 10. Take this 1 and add it to the 7, and this makes 8, which is the remainder of this order and is written underneath the 9. Then carry 1 to the 5, and this makes 6, and 6 from 9 is 3. Write this 3 under the 5, and the whole remainder is 38. Thus we have proved in two ways that the first sum which we found in the chapter on addition was correct. Similarly with the present operation of subtraction; it can be proved by this sum, as you have been allowed to do in this chapter. And since all of the examples in the chapter on addition can be proved by the operation of subtraction, I do not wish to give you in this chapter other examples, except by reference to all those which you will find in that chapter. Furthermore, by the operation of addition you are able to prove all the operations of the present chapter. And other examples will not be given in this chapter save indirectly, that you take the examples of that chapter in order, and subtract one of the numbers from the sum and the result will be the other number, and once is enough.

F. 11, v.

Take, then, the second example which you found in that chapter, and find the result appropriate to this chapter.

Wishing to take 816 from 2732, arrange your work as follows:

	2732
	816
The remainder	1916

Then begin to subtract thus: 6 from 2 is impossible; but 4 is the complement with respect to 10, and this 4 and 2 make 6, which is the remainder. Write this 6 below. Then having used a complement, carry 1 to the other 1, making 2, and subtract this 2 from 3, leaving 1. Write this 1 beneath the other 1. Then 8 cannot be subtracted from 7, but 2 is the complement of 8, and this 2 added to 7 gives 9 for the remainder, which is written below. Then carry 1 to the vacant place and take it from 2, 1 from 2 leaving 1, which is written underneath the vacant place, and the result is 1916. I will now leave the method of proof of this example and all succeeding ones to your care, since if you have an interest in study you will understand clearly the method and manner from what has already been given.

Let us now consider a third example, arranging the work as follows. Let us subtract 2732 from 48050, thus:

	48050
	2732
The remainder	45318

Now begin to subtract in this way: It is impossible to subtract 2 from 0,

F. 12, r.

so we take 8, the complement of 2, and write this 8 beneath the 2 for a remainder, and carry 1 to the 3, making 4. Then 4 from 5 leaves 1, which 1 we write beneath the 3. Since we cannot take 7 from 0, we take 3, the complement, and write it beneath the 7, and carry 1 to the 2, minus 3. This 3 we take from 8, leaving 5, and this we write underneath the 2. In the vacant place we take 0 from 4, leaving 4, and this we write below the vacant place. The result is then 45318. Now prove that this is correct according to the method which I have given.

Take the fourth example and arrange your work as follows: Let us subtract 392 lire from 961 lire, thus:

	lire	961
	lire	392
The remainder	*lire*	569

Begin with 2 from 1, but since this subtraction is impossible, take 8, the complement of 2, and add this 8 to 1, making 9, which write beneath 2 as the remainder. Then carry 1 to 9, making 10. Since 10 has no complement you have only to write as a remainder the 6 which is above the 9. Then carry 1 to the 3, making 4, and subtract this 4 from 9, leaving 5, which you write underneath the 3. Thus the remainder is 569.

To understand the fundamental principles, it should be observed that the reasoning involved in this chapter comes down simply to this: that 10 is the radix of all abstract numbers, including those that relate to the larger weights, or measures, or monies. So when we write lire, soldi, grossi, and pizoli as a single sum, the radix of the lire, the largest denomination, is 10; and that of the soldi is 20, since 20 soldi make a lira; and that of the grossi is 12, since 12 grossi make a soldo; and that of the pizoli is 32, since 32 pizoli make a grosso. And thus we can understand the radices of all other money, or weights, or measures, according to the number of the smaller units in the larger one.

F. 12, v.

Consider now the fifth example, involving lire and soldi, and note that the radix of the soldi is 20, and the radix of the lire (as I have explained above) is 10. Arrange the work of the present proposition as follows: If you are required to take 1945 lire 15 soldi from 2862 lire 9 soldi, you will proceed thus:

	lire	2862	s. 9
	lire	1945	s. 15
The remainder	lire	916	s. 14

Beginning to subtract, we cannot take 15 from 9; but 5 is the complement of 15 with respect to 20. Add this 5 to the 9, making 14, and this is the remainder for the soldi, which is now written beneath the 15. Then carry the 1 to the 5, making 6; 6 cannot be subtracted from 2; 4 is the complement of 6; 4 and 2 make 6, and this is the remainder to be written beneath the 5. Then carry 1 to 4, making 5; take 5 from 6, and write 1 beneath the 4. We cannot take 9 from 8; but 1, the complement of 9, added to 8, makes 9, which is written as the remainder beneath 9. Then carry 1 to the other 1, making 2, and subtract 2 from 2, leaving 0. Therefore we write nothing, because whenever there are no more figures to the left we stop, there being nothing more to do. Thus the remainder is 916 lire, 14 soldi.

F. 13, r.

Let us now consider example six, involving lire, soldi, and pizoli. Noting also that the radix of pizoli is 12, proceed with the problem. If you are asked to subtract 9562 lire, 19 soldi, 11 pizoli from 10455 lire, 15 soldi, 6 pizoli, arrange the work as follows:

	lire	10455	s. 15	p. 6
	lire	9562	s. 19	p. 11
The remainder	lire	892	s. 15	p. 7

We begin the subtraction by noting that it is impossible to take 11 from 6. The complement of 11 with respect to 12 is 1, which we add to the 6 making 7, which we write below the 11 as the difference of the pizoli. Then we carry 1 to the 19, making 20. And since we cannot add anything to 20 as a complement, we write the 15, which is above the 19, below it as a remainder. We then carry 1 to the 2, making 3, and subtract 3 from 5, leaving 2, which we write as a remainder below the 2. Since we cannot take 6 from 5 we add the complement 4 to 5, making 9, and write this under the 6. We then carry 1 to the 5, making 6, and since we cannot take 6 from 4 we add 4, the complement of 6, to the other 4, making 8, which we write below the 5 as a remainder. We then carry 1 to the 9, making 10, of which the complement is 0, as above, which we write as a remainder below the 9. We then carry 1 to the vacant place, and 1 from 1 leaves 0, which we do not write for the reason above given. Then the remainder is 892 lire, 15 soldi, 7 pizoli.

Now let us consider the seventh example, which relates to ducats, grossi, and pizoli. Therefore it must be known that the radix of the pizoli is 32, of the grossi 24, and of the ducats 10, and we proceed with the problem. If you are asked to subtract 2169 ducats, 23 grossi, 31 soldi from 4072 ducats, 16 grossi, 22 pizoli, arrange the work as follows:

	duc.	4072	g. 16	p. 22
	duc.	2169	g. 23	p. 31
The remainder	duc.	1902	g. 16	p. 23

F. 13, v.

We begin the subtraction thus: we cannot take 31 from 22, but 1 is the complement of 31 with respect to 32. We add this 1 to 22, making 23, which we write as a remainder below the 31 pizoli. We then carry 1 to 23, making 24, which is its own radix. We therefore write below the 23 the 16 which is above it, and this is the remainder of the grossi. We then carry 1 to the 9, making 10, which is also its own radix. We therefore write below the 9 the 2 which is above it, and this is the remainder. We then carry 1 to 6, making 7, and take 7 from 7, leaving 0, which we write below the 6. Since we cannot take 1 from 0, we write its complement 9 below the 1 as the remainder. We then carry 1 to 2, making 3, and take 3 from 4, leaving 1, which we write as a remainder beneath the 2. Thus we have for the entire remainder 1902 ducats, 16 grossi, 23 pizoli.

Let us now consider the eighth example, as follows: If you are asked to take 2820 lire, 5 soldi, 4 grossi, 8 pizoli, from 8433 lire, 4 soldi, 3 grossi, 27 pizoli, of which the radices have already been given, arrange your work as follows:

	lire	8433	s. 4	g. 3	p. 27
	lire	2820	s. 5	g. 4	p. 8
The remainder	lire	5612	s. 18	g. 11	p. 19

Now begin taking 8 from 27, leaving 19, which is written beneath the 8 as the difference of the pizoli. Then since we cannot take 4 from 3 we add 8, the complement of 4 with respect to 12, to the 3, making 11, which we write as the difference of the grossi below the 4. We then carry 1 to the 5, making 6. Since we cannot take 6 from 4, we add 14, the complement of 6 with respect to 20, to the 4, making 18, which we write as the difference of the soldi below the 5. We then carry 1 to the 0 place, and take 1 from 3, leaving 2, which we write as a remainder below the 0.

F. 14, r.

Then we take 2 from 3, leaving 1, which we write as a remainder below the 2. Since we cannot take 8 from 4 we add 2, the complement of 8 with respect to 10, to 4, making 6, which we write as a remainder below the 8. We then carry 1 to 2, making 3, and take 3 from 8, leaving 5, which we write as a remainder below the 2. The total remainder is then 5612 lire, 18 soldi, 11 grossi, 19 pizoli. From the above examples we are able to understand the method of subtracting one quantity from another, what are the radices described, and also the necessity of considering other radices to know how to find the complements. And this sufficeth as to the operation of subtraction.

Having now explained the third operation, namely that of subtraction, the reader should give attention to the fourth, namely that of multiplication. To understand this it is necessary to know that to multiply one number by itself or by another is to find from two given numbers a third number which contains one of these numbers as many times as there are units in the other. For example, 2 times 4 are 8, and 8 contains 4 as many times as there are units in 2, so that 8 contains 4 in itself twice. Also the 8 contains 2 as many times as there are units in 4, and 4 has in itself four units, so that 8 contains 2 four times. It should be well understood that in multiplication two numbers are necessary, namely the multiplying number and the number multiplied, and also the multiplying number may itself be the number multiplied, and vice versa, the result being the same in both cases. Nevertheless usage and practice demand that the smaller number shall be taken as the

F. 14, v.

multiplying number, and not the larger. Thus we should say, 2 times 4 makes 8, and not 4 times 2 makes 8, although the results are the same. Now not to speak at too great length I say in brief, but sufficiently for the purpose of a Practica, that there are three methods of multiplication, namely by the tables, cross multiplication, and the chessboard plan. These three methods I will explain to you as briefly as I am able. But before I give you a rule or any method, it is necessary that you commit to memory the following statements, without which no one can understand all of this operation of multiplication. Therefore learn:

F. 15, r.

2	times	2	makes	4
2	times	3	makes	6
2	times	4	makes	8
2	times	5	makes	10
2	times	6	makes	12
2	times	7	makes	14
2	times	8	makes	16
2	times	9	makes	18
2	times	0	makes	0
3	times	3	makes	9
3	times	4	makes	12
3	times	5	makes	15
3	times	6	makes	18
3	times	7	makes	21
3	times	8	makes	24
3	times	9	makes	27
3	times	0	makes	0
4	times	4	makes	16
4	times	5	makes	20
4	times	6	makes	24
4	times	7	makes	28
4	times	8	makes	32
4	times	9	makes	36
4	times	0	makes	0
5	times	5	makes	25
5	times	6	makes	30
5	times	7	makes	35
5	times	8	makes	40
5	times	9	makes	45
5	times	0	makes	0

6	times	6	makes	36
6	times	7	makes	42
6	times	8	makes	48
6	times	9	makes	54
6	times	0	makes	0
7	times	7	makes	49
7	times	8	makes	56
7	times	9	makes	63
7	times	0	makes	0
8	times	8	makes	64
8	times	9	makes	72
8	times	0	makes	0
9	times	9	makes	81
9	times	0	makes	0

F. 15, v.

Those who are of scholarly tastes should learn by heart the following table, since it is necessary for all who wish to make use of the operation of multiplication. Others, too, should learn it, although it is not necessary, save as it relates to the radices of the tables of money, measure, and weight.

To reduce soldi to pizoli, and pounds to ounces.

1	time	12	makes	12
2	times	12	makes	24
3	times	12	makes	36
4	times	12	makes	48
5	times	12	makes	60
6	times	12	makes	72
7	times	12	makes	84
8	times	12	makes	96
9	times	12	makes	108
0	times	12	makes	0

To reduce lire to soldi.

1	time	20	makes	20
2	times	20	makes	40

3	times	20	makes	60
4	times	20	makes	80
5	times	20	makes	i00
6	times	20	makes	i20
7	times	20	makes	i40
8	times	20	makes	i60
9	times	20	makes	i80
0	times	20	makes	0

F. 16, r.

To reduce ducats to grossi, in gold.

i	time	24	makes	24
2	times	24	makes	48
3	times	24	makes	72
4	times	24	makes	96
5	times	24	makes	i20
6	times	24	makes	i44
7	times	24	makes	i68
8	times	24	makes	i92
9	times	24	makes	2i6
0	times	24	makes	0

To reduce grossi to pizoli, in gold.

i	time	32	makes	32
2	times	32	makes	64
3	times	32	makes	96
4	times	32	makes	i28
5	times	32	makes	i60
6	times	32	makes	i92
7	times	32	makes	224
8	times	32	makes	256
9	times	32	makes	288
0	times	32	makes	0

To reduce quarti to carats.

i	time	36	makes	36
2	times	36	makes	72
3	times	36	makes	i08
4	times	36	makes	i44
5	times	36	makes	i80

F. 16, v.

6	times	36	makes	216
7	times	36	makes	252
8	times	36	makes	288
9	times	36	makes	324
0	times	36	makes	0

I have now given you to learn by heart all the statements needed in the Practica of arithmetic, without which no one is able to master the Art. We should not complain, however, at having to learn these things by heart in order to acquire readiness; for I assure you that these things which I have set forth are necessary to anyone who would be proficient in this art, and no one can get along with less. Those facts which are to be learned besides these are valuable, but they are not necessary.

Having learned by heart all of the above facts, the pupil may with zeal begin to multiply by the table. This operation arises when the multiplier is a simple number, and the number multiplied has at least two figures, but as many more as we wish. And that we may more easily understand this operation we shall call the first figure, toward the right, units; the second, toward the left, tens, and the third shall be called hundreds. This being understood, attend to the rule of working by the table, which is as follows: First multiply together the units of the multiplier and the number multiplied. If from this multiplication you get a simple number, write it under its proper place; if an article, write a 0 and reserve the tens to add to the product of the tens; but if a mixed number is found, write its units in the proper place, and save the tens to add to the product of the tens, proceeding in the same way with all the other orders. Then multiply together the units of the multiplier with the tens; then with

F. 17, r.

the hundreds, and so on in regular order. Now let us consider an example.

Suppose you are asked to find the product of 8 times 9279, proceed as follows: Multiply 8 times 9, making 72; write 2 and carry 7. Then 7 times 8 are 56, and 7 which was reserved makes 63; write 3 and carry 6. Then 2 times 8 are 16, and 6 reserved are 22; write 2 and carry 2. Then multiply 8 times 9, making 72, and 2 reserved makes 74; write first 4 and then 7 to the left, and the total result is 74232.

If you wish to prove this result by casting out 9s, add all the digits of the multiplier and of the number multiplied, but neglect every 9 or 0 which you find, since the excess in every 9 is 0, and that in every 0 is 0, so subtract 9 whenever you find it, retaining the rest. So if you would prove the above multiplication, arrange the work thus:

9279	0
8	8
74232	0

Now begin to find the excess, and the excess in 9279 is 0, thus: 2 and 7 are 9, which we may call 0. So the excess in 9279 is 0, and this 0 we write after the number and beyond the bar. The excess in 8 is 8, which we also write after the other 8 and beyond the bar. Now multiply these two excesses, the one by the other, and 8 times 0 is 0, which you should write after the result. Now see if the result has also this excess 0, thus: 7 and 4 are 11; subtract 9 and 2 remains; then this 2 and the other 2 are 4, and 3 are 7, and 2 are 9, and the excess of 9 is 0. Thus the result is correct, and this method of proof applies in all other cases.

F. 17, v.

There is, however, a better proof for this form of multiplication. If you divide 74232 by 8 there will be found for a quotient 9279; or if you divide 74232 by 9279, the quotient will be 8. But this method of proof I cannot give you until you understand division. Consequently division proves multiplication and multiplication proves division. This method of proof will be fully discussed in teaching the operation of division, in the second example in that chapter.

If you are asked to find 7 times 12392, proceed thus: Multiply 7 with each of the digits of the larger number, beginning with the first to the right, that is with 2, then taking the others in their order to the left. Beginning, then, we have 2 times 7 are 14; write the 4 and carry 1. Then 7 times 9 are 63, and 1 to carry makes 64; write 4 and carry 6. Then 3 times 7 are 21, and 6 to carry makes 27; write 7 and carry 2. Then 2 times 7 are 14, and 2 to carry makes 16; write 6 and carry 1. Then 1 times 7 is 7, and 1 to carry makes 8, which is written in this place. The result of 7 times 12362 [12392] is, therefore, 86744.

If you wish to prove this arrange the work in the following form:

12392	8
7	7
86744	2

We now prove the result thus: 1 and 2 are 3, and 3 are 6, and 2
are 8, so that the excess of 12392 is 8, which is placed after the number, separated by the bar. Then we see that the excess of 7 is 7, and this is placed after the bar, following the other 7.

F. 18, r.

Now multiply these two excesses together, thus: 7 times 8 are 56, and 5 and 6 are 11; subtract 9 and 2 remains as the principal excess. Now see if the excess of 86744 also equals 2, thus: 8 and 6 are 14; subtract 9 and 5 remains; 5 and 7 are 12; subtract 9 and 3 remains; then 3 and 4 are 7, and 4 are 11; subtract 9 and 2 remains. Thus the result is correct, and by this method it is possible to prove all results of this kind.

Since you now understand the first method of multiplication, i.e., by the table, attend diligently to the second method, i.e., simple cross multiplication. This method is employed when you have to multiply one number of two figures by another of two figures, as in the case of 12 times 12. This is the rule. First multiply units by units and of the result of the multiplication write the units and carry the tens. Then multiply together the units of one number and the tens of the other, crosswise, adding the tens carried from the first multiplication, and of this result write here the units and carry the tens. Then multiply the tens by the tens, adding the tens which you carried, and write the result and the work is complete.

If you are required to find 12 times 13, do as follows: Multiply 2 times 3, making 6, and write this 6 under the units, carrying nothing because there is no other figure. Then multiply crosswise, thus: 1 times 3 is 3, and 1 times 2 is 2; add 2 and 3, making 5, and write this 5 under the tens. Then multiply

F. 18, v.

1 time 1, making 1, writing this 1 at the left of the 5, and the result is 156.

If you wish to prove the above by casting out 9s, write the work of cross multiplication as follows:

13	4
12	3
156	3

Then the excesses of the numbers are found as follows: for 13 we have 1 and 3 are 4, which we write after the bar; then for 12 we have 1 and 2 are 3, which we also write after this number, separated by the bar. Then we multiply one excess by the other, thus: 3 times 4 are 12; subtract 9 and 3 remains. So the principal excess is 3, which we place after 156, separated by the bar. Now see if the excess of 156 is also 3, thus: 1 and 5 are 6, and 6 are 12; subtract 9 and 3 remains, and thus we see that the result is correct. In the same way we can prove all other examples in cross multiplication.

If you are asked how much is 48 times 56, proceed as follows: multiply 6 times 8, making 48; write 8 under the units and reserve 4. Then multiply crosswise, thus: 4 times 6 are 24, and 5 times 8 are 40; add 24 and 40, giving 64, and add the 4 which was carried, making 68; write 8 and carry 6. Now multiply the tens by the tens, thus: 4 times 5 are 20, and 6 to carry are 26, which is written in its proper place. The result is therefore 2688. We then say that 48 times 56 are 2688. In this same way you can perform all other cross multiplications.

If you wish to prove this result, arrange the work as follows:

F. 19, r.

56	2
48	3
2688	6

Then take the excess of 56, thus: 5 and 6 are 11; subtract 9 and 2 remains; place this 2 after the 56, separated by the bar. Then for 48: 4 and 8 are 12; subtract 9 and 3 remains; place this 3 after the 48, separated by the bar. Then multiply these excesses, the one by the other, thus: 2 times 3 are 6; place this 6 after the 2688, separated by the bar. Now see if the excess of 2688 is also 6, thus: 2 and 6 are 8, and 8 are 16, and 8 are 24; adding the digits of 24, 2 and 4 are 6, which is the excess in 2688, which shows that the result is correct. By this same plan you can prove all other examples in cross multiplication.

If you are asked what is the product of 85 and 98, proceed as follows: 5 times 8 are 40; write the 0 under the units and carry 4. Then multiply crosswise, thus: 8 times 8 are 64, and 5 times 9 are 45; add 64 and 45, making 109, and 4 which was carried, making 113; write 3 and carry 11. Then 8 times 9 are 72, and 11 which was carried makes 83; write all this to the left, giving 8330. Therefore, the result of 85 times 98 is 8330.

If you wish to prove this, arrange your example in cross multiplication as follows:

98	8
85	4
8330	5

F. 19, v.

Now begin the proof thus: The excess of 98 is 8, since 9 does not count. Therefore write 8 after the other 8, separated by the bar. Then for 85, 8 and 5 are 13; subtracting 6 [9], 4 remains, and this 4 is written after the 5, separated by the bar. Now multiply these two excesses, one by the other, thus: 4 times 8 are 32, the sum of the digits of which 32, 3 and 2, is 5, which is written after the 0, separated by the bar. Now see if the excess in 8330 is also 5, in which case the result is correct. We have 8 and 3 are 11; subtracting 9, 2 remains; add this to the other 3 and we have 5, for the 0 (as I have stated before) does not count. Thus the result is correct. In the same way you can prove all other examples in cross multiplication.

Having now explained cross multiplication, which is the second method, let us consider the third, namely, that of the chessboard, which is used when we have to multiply together a number of at least two figures and one of three or more, according to the circumstances. And note carefully that if you know thoroughly multiplication by the table, you will also know multiplication by the chessboard method, for you have nothing else to do but to multiply as in the table by each figure of the multiplier, writing each product under its own multiplier and proceeding to the left, as you will see by an example.

If you are asked to find 24 times 829, multiply first by 4, thus: 4 times 9 are 36; write 6 and carry 3; then 2 times 4 are 8, and 3 which was carried are 11; write 1 and carry 1; then 4 times 8 are 32, and 1 which was carried are 33, which is all written in its place. The result of this part of the multiplication is 3316, which is written down in the following form:

F. 20, r.

```
 829|
  24|
----|
3316|
```

Now perform the other multiplication, i.e., multiply 829 by 2, thus: 2 times 9 are 18; write 8 under the 2, that is in tens' place, and carry 1; then 2 times 2 are 4, and the 1 carried are 5;

write this 5 to the left; then 2 times 8 are 16; write the entire 16 to the left, and the work with both multiplications appears as follows:

```
    829|
     24|
  -----|
   3316|
  1658 |
```

Now add these two products in the manner already learned, the sum being 19896. If you wish to prove this work, arrange it as follows:

```
    829 | 1
     24 | 6
  ------|---
   3316 |
  1658  |
  ------|---
  19896 | 6
```

Now find the excess in 829, thus: 8 and 2 are 10, cancelling the 0, 1 remains, and this 1 is written after the number, separated by the bar. Then find the excess in 24, thus: 2 and 4 are 6; write this 6 after its number, separated by the bar. Then multiply one excess by the other, thus: 1 time 6 is 6, and write this 6 after the sum of the two products, separated by the bar. If now the excess of the sum also amounts to 6 the work is correct. Therefore, we find that 1 and 8 are 9, which has 0 excess, and that 6 remains as the excess. Hence the result is correct, and this method can be used in all cases of the chessboard multiplication.

F. 20, v.

If you are asked the product of 314 times 934, arrange the work as follows:

```
934
314
```

Now proceed thus: find the first product, i.e., 4 time 934. Here we have 4 times 4 are 16; write 6 and carry 1; then 3 times 4 are 12, and 1 carried is 13; write 3 and carry 1; then 4 times 9 are 36, and 1 carried makes 37, all of which is now written, so that the first multiplication gives 3736 as you see here:

```
  934 |
  314 |
 ---- |
 3736 |
```

Now find the second product, namely, 1 time 934, which is 934, since the order of the multiplier does not change the figures. Then write these two products as here shown:

```
  934 |
  314 |
 ---- |
 3736 |
 934  |
```

Now find the third product: 3 times 934, thus: 3 times 4 are 12; write 2 in its place, i.e., in the hundreds' place, and carry 1; then 3 times 3 are 9, and 1 carried makes 10; write 0 in its place to the left, and carry 1; then 3 times 9 are 27, and 1 carried makes 28; write 28 in its place to the left. This finishes the multiplication, and we write all three of the products in the following form:

F. 21, r.

```
    934 | 7
    314 | 8
 -------|---
   3736 |
   934  |
 2802   |
 -------|---
 293276 | 2
```

Now find the sum of these products, thus: 6, which is 6 written directly under that 6, below the bar; then 3 and 4 are 7, and we write this 7 in its proper place; 2 and 3 are 5, and 7 are 12; write 2 and carry 1; then this 1 added to 9 makes 10, and 3 are 13; write 3 underneath in its proper place and carry 1; then 8 and 1 are 9; write 9 under 8; then take the 2 and write it below in its proper place, and the work is done. Therefore the result of 314 times 934 is 293276.

If we wish to prove this work called the chessboard multiplication, we take first the excess in 934, thus: 3 and 4 are 7, which 7 we write after the number, separated by the bar. Then take the excess in 314, thus: 3 and 1 are 4, and 4 are 8, which 8 we also write after the number, separated by the bar. Then multiply one excess by the other, thus: 7 times 8 are 56; adding the digits, 5 and 6 are 11; subtracting 9, 2 remains as the principal excess, and this is written below, after the result, separated by the bar. Now see if the excess of the result is also 2, thus: 2 and 3 are 5, and 2 are 7 are 14; subtracting 9, 5 remains; adding 6 we have 11; subtracting 9, 2 remains, and thus the result is correct. In the same way we are able to proceed with all examples of this kind.

And this is sufficient for your information concerning the three methods of multiplying.

F. 21, v.

I wish, however, that you may know other forms of chessboard multiplication, but I shall leave them for your own investigation, merely setting forth the forms as you see them below.

Now take the example given above in the chessboard method, 314 times 934, and note the four methods given below.

```
      9 3 4
    3 7 3 6   4
    9 3 4     i
2 8 0 2   3
2 9 3 2 7 6
```

		9	3	4		
		3	7	3	6	4
			9	3	4	i
		2	8	0	2	3
2	9	3	2	7	6	

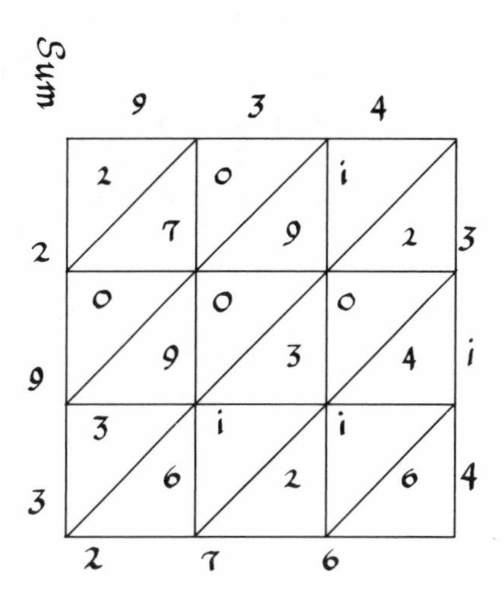

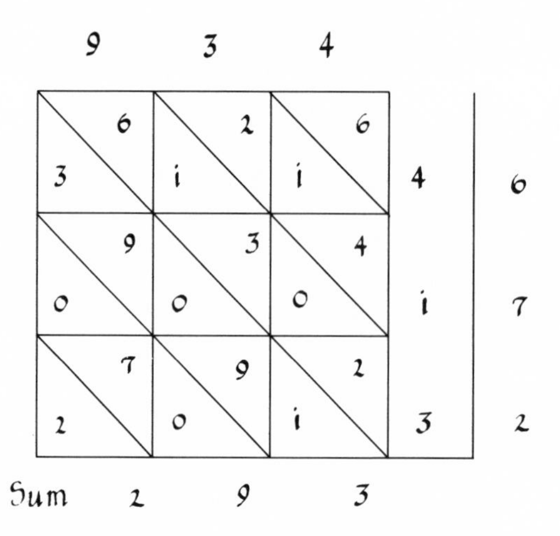

F. 22, r.

If you are asked how much is 1234 times 56789, proceed by the five methods below.

```
             5 6 7 8 9
         2 2 7 1 5 6  / 4
       1 7 0 3 6 7  / 3
     1 1 3 5 7 8  / 2
     5 6 7 8 9  / 1
Sum  7 0 0 7 7 6 2 6
```

```
          5 6 7 8 9
            i 2 3 4
      -------------
        2 2 7 i 5 6
      i 7 0 3 6 7
    i i 3 5 7 8
    5 6 7 8 9
    ---------------
Sum 7 0 0 7 7 6 2 6
```

	5	6	7	8	9		
2	2	7	i	5	6	4	
i	7	0	3	6	7	3	6
i	i	3	5	7	8	2	2
	5	6	7	8	9	i	6
Sum	7	0	0	7	7		

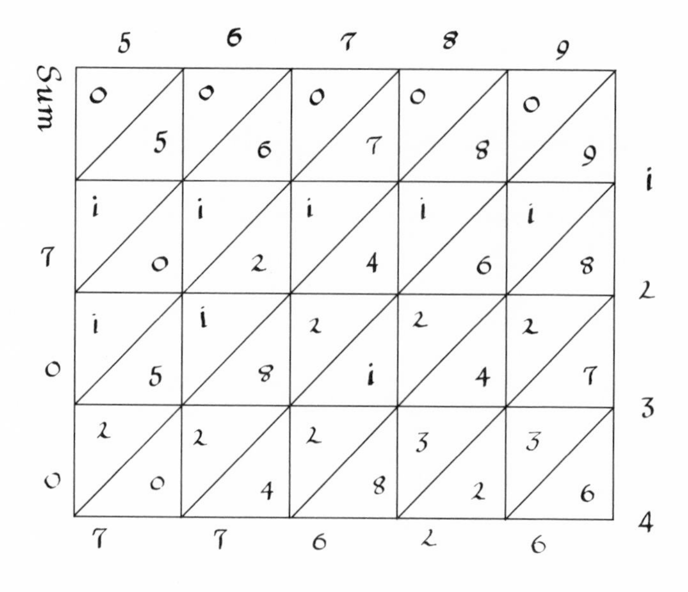

F. 22, v.

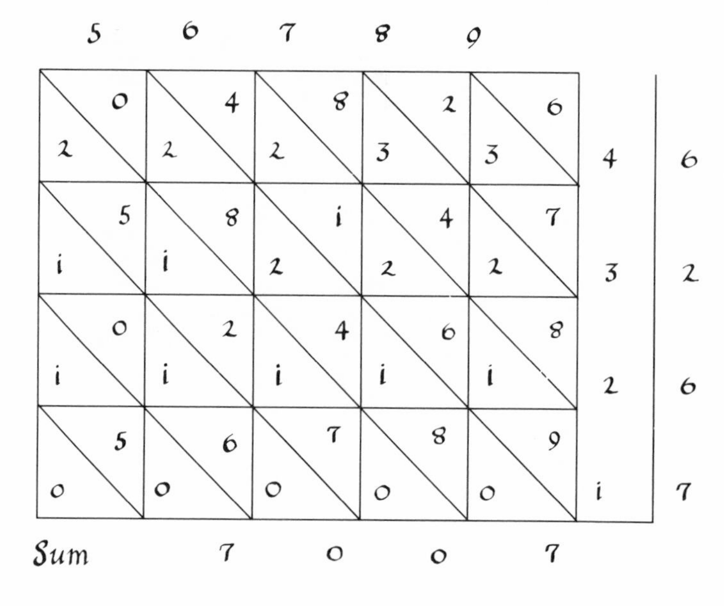

By the above five methods you can solve any problem in chessboard multiplication, but it is better to use the one which I stated as the easiest for you. And this sufficeth for the third operation.

In order to understand the fourth operation, i.e., division, three things are to be observed, i.e., what is meant by division; second, how many numbers are necessary in division; third, which of these numbers is the greater. As to the first I say that division is the operation of finding, from two given numbers, a third number, which is contained as many times in the greater number as unity is contained in the less number. You will find this number when you see how many times the less number is contained in the greater. Suppose, for example, we have to divide 8 by 2; here 2 is contained 4 times in 8, so we say that 4 is the quotient demanded. Also, divide 8 by 4. Here the 4 is contained 2 times in 8, so that 2 is the quotient demanded.

Second, it is to be noticed that three numbers are necessary in division, the number to be divided, the divisor, and the quotient, as you have understood from the example above given, where 2 is the divisor, 8 the number to be divided, and 4 the quotient. From this is derived the knowledge

F. 23, r.

of the third thing which is to be noted, that the number which is to be divided is always greater than, or at least is equal to, the divisor. When the numbers are equal the quotient is always 1.

Now to speak briefly, it is sufficient in practice to say that there are two ways of dividing, by the table and the galley method. In this operation you should begin with the figure of highest value, that is by that which is found at the left, proceeding thence to the right. If you can divide by the table you will be able to divide by the galley method, and it is well, for brevity, to avoid the latter when you can. Therefore this is the method of dividing by the table: See how many times your divisor is found in the first left-hand figure, if it is contained in it, and write the quotient beneath it. If it is not so contained, consider this figure as tens and take together with it the following figure; then, finding the quotient write it beneath the smaller of the two figures. If there is any remainder, consider this as tens, and add it to the next number to the right, and see how many times your divisor is found in these two figures, writing the quotient under the units. In this same way proceed with the rest of the figures to the right. And when you have exhausted them all, having set down the quotient, write the remainder at the right, separated by a bar; and if the remainder is 0, place it where I have said. In the name of God I propose the first example, so attend well.

Divide 7624 ducats into two parts, i.e., by 2, arranging your work as follows:

F. 23, v.

The divisor	.2.	7624	0	The remainder
The quotient		3812		

Now consider how many times your divisor, 2, is found in 7, thus: 2 is found 3 times in 7; write 3 under the 7; the remainder, 1, which I is left in its place, taken together with 6, makes 16; then 2 is found 8 times in 16, and this is written under the 6; then 2 in 2 is 1, and this 1 is written under 2; then 2 in 4 is 2, which is written under 4, and the 0 remainder is written after the 4, separated by the bar. Thus the quotient in 3812.

If you wish to prove this by the best proof, multiply the quotient by the divisor, and if the result is the number divided the work is correct.

If you wish to prove it by casting out 9s, put the excess in the divisor, which is 2, in a little cross, underneath the left; then put the excess in the quotient, which is 5, above this 2; then place the excess in the remainder, which is 0, after the 5 on the other side. Then do as follows: multiply the excess of the divisor by that of the quotient, 2 times 5 making 10; add the 0 remainder, leaving 10; cancel the 0, cleaving 1 for the principal excess, and write this in the cross under the excess of the remainder. Then see if the excess of the number divided also equals 1, in which case the result is correct. To understand the permission given in the chapter on multiplication, to prove one operation by the other, multiplication, by division and division by multiplication, turn to that chapter, taking the examples in their order and using them in the present case. Take therefore the

5	0
2	1

first one found in multiplication by the table, and arrange it as follows:

```
 9279|
    8|
-----|
74232|
```

F. 24, r.

Now divide 74232 by 8 and the quotient will be 9279, so that it proves. So one operation is proved by the other. Now arrange the work in the following form.

The divisor	.8.	74232	0
The quotient		9279	The remainder

Now take your divisor, which is 8, and say: "8 is not contained in 7." Take, therefore, 7 and 4 together. 8 is contained 9 times in 74, so write 9 under 4. The remainder is 2, since 8 times 9 is 72, and you have 74. So you have a remainder 2, which 2 is represented in its place, that is in the place of 4, and with the other 2 following makes 22. Then 8 in 22 is 2, which is written under 2, and the remainder is 6, for 2 times 8 is 16, and you have 22, so that there is 6 over. This 6, considering its place, together with the 3, makes 63. Then 8 in 63 is 7, which is written under 3, and the remainder is 7, for 7 times 8 is 56, and this from 63 leaves 7. This 7, considering its place, together with the 2 following, makes 72. Then 8 in 72 is 9, which is written under 2, and the remainder is 0, which is written after 2, separated by the bar. The quotient is, therefore, 9279. This is proved by multiplication, as the latter is proved by this, as I promised in that chapter. We may also prove by excess of 9s that the quotient is correct, as follows: Place the excess in the divisor, namely, 8, in the cross, to the left and below; then place the excess in the quotient, 0, above the 8; then place to the right and above the excess in the remainder, which is 0. Having done this, multiply thus: 8 times 0 is 0, and since the remainder is 0 it follows that 0 is the principal excess, and this is written underneath the excess of the remainder. Now if the excess of the dividend, 74232, is also 0, the result is correct.

0	0
8	0

F. 24, v.

Take now the second example in multiplication, by column, and arrange it for the present case.

Divide 86744 by 7. Then if the quotient is 12392, it will be correct. Arrange the work as follows:

The divisor	.7.	86744	0 *The remainder*
The quotient		12392	

Begin the division by taking the divisor, 7, thus: 7 in 8 gives 1; write 1 under 8 for the quotient, and the remainder is 1. This 1, considering its place, together with the 6 that follows, makes 16. Then 7 in 16 gives 2, which is written beneath 6, and the remainder is 2. This 2, considering its place, together with the 7 that follows, makes 27. Then 7 in 27 gives 3, which is written beneath 7, and the remainder is 6. This 6, considering its place, together with the 4 that follows, makes 64. Then 7 in 64 gives 9, which is written beneath and the remainder is 1. This, considering its place, together with the 4 that follows, makes 14. Then 7 in 14 gives 2, which is written beneath the 4, and the remainder, 0, is written after the 4, separated by the bar. This therefore proves that the table multiplication was correct, and this operation of division is also proved by that of multiplication. These two preceding examples suffice to show that the operation of multiplication proves that of division and that of division proves that of multiplication. In the same way it is possible to prove all other results.

If you wish to prove the work by casting out 9s, place the excess of the divisor in the cross, underneath to the left, that is 7. Then place the excess of the quotient, 8, above the 7. Then place on the other side of the cross and above, the excess of the remainder, which is 0. Then multiply the

F. 25, r.

excess in the quotient by that in the divisor, thus: 7 times 8 are 56; add the remainder, namely 0, and you still have 56. Adding the digits, 5 and 6 are 11, and subtracting 9 there remains 2 as the principal excess. Write this under the excess of the remainder. Then see if the excess of the result is also 2, in which case the work is correct.

8	0
7	2

And this sufficeth for the understanding of division by the table.

Having now learned the first method of division, viz., by the table, attend to the second method, which is a trifle more difficult. Therefore, before entering into this work it is very necessary that you be certain that the preceding method is understood, by finding exactly the quotient, and at the same time that you are rapid and accurate in the operation of multiplication and in the operation of subtraction, for all of these operations enter into this one. I tell you in advance, however, that the two examples which I shall give in division by the galley method admit also of solution by the table, but they are given as being more easily understood. Attend, then, diligently to the arrangement of this operation that I set forth in the first example, as follows:

If you wish to divide 825 by 2 arrange the work like this:

```
8 2 5 | 4
2     |
```

Placing your divisor, 2, below 8, note carefully how many times 2 is contained in 8, and this is 4, which is the quotient derived from 8. Write this after the 5 and beyond the bar, where the quotient belongs. Then proceed as follows: multiply the quotient, 4, by the divisor, 2, thus: 2 times 4 are 8. Now, not writing this 8, but keeping in mind what came from the operation of multiplying, i.e., 8, cancel this 8, which you kept in mind, from the other 8 which was above the 2, saying "8 from 8," and crossing it out with the pen, the remainder being 0. This, then, is the order: first to place the divisor under the first figure; then to note the quotient and write it in its proper place; then to multiply the quotient by the divisor; then to subtract the figure produced by the multiplication from the figure of the number to be divided. This order you must preserve.

F. 25, v.

Now to continue: arrange the work beneath the preceding work in the following manner:

```
8̸ 2 5 | 4 1
2̸ 2   |
```

Now proceed as follows: place your divisor, 2, under that 2 which follows the 8. Then see how many times this 2 is contained in the other 2, saying: "2 in 2 gives 1." Write this in the quotient to the right of the 4. Then do this: Multiply that quotient, 1, by the divisor, 2, saying: "1 time 2 makes 2." Now not writing this 2, subtract it from the 2 which follows the 8, saying: "2 from 2," and cancelling this 2 with the pen, the remainder being 0. Now to continue: arrange the work beneath the preceding work in the following manner:

825 | 412
222 |

Now proceed as follows: place your divisor, 2, under the 5. Then say: "2 in 5 gives 2," place this 2 as the quotient

F. 26, r.

after 41, multiply this 2 of the quotient by the 2 of the divisor, which is below 5, saying: "2 times 2 are 4." Do not write this 4, but subtract it from 5, saying: "4 from 5," cancelling the 5 and writing the remainder, 1, above the 5. And now observe that whenever, in subtracting one figure from another, you have a remainder, this should be written above the figure from which it was obtained, as here.

1
825 | 412
222 |

And thus is completed your work. Therefore the quotient of 825 by 2 is 412.

And proceeding in this way we may perform any long division by the galley method, however many figures there may be in the number to be divided.

Now give attention. In this galley division you need to know four terms which enter, each by its own name. The first is the number divided, that is 825. The second is the divisor, 2. The third is the quotient, 412. The fourth is the remainder, 1. And in this way it is necessary that you recognize the four terms which will be found in every other example in galley division, so that you shall be able to prove that the work is correct.

If, then, you wish to prove the above result of galley division, make a cross in which to place the excess of each number. Now if you would prove by the excess of 9s, take the excess in the divisor, 2, and place it in the cross, below and to the left. Then take the excess of

F. 26, v.

the quotient, that is of 412, which is 7, and place it above the 2. Then place the excess of the remainder, which is 1, after the 7. Having thus arranged the three excesses, multiply that of the divisor by that of the quotient, saying: "2 times 7 are 14," and add the excess of the remainder, 1, and this will be 15; then subtract 9, and 6 remains, which is the principal excess, and this is written below 1. Now see if the excess in the number divided is also 6, in which case the work is correct. If you wish to prove the operation by the operation of multiplication, multiply (as I have said before) the quotient by the divisor, and add the remainder; if the result is the number divided, it is correct. And you can prove all other cases of galley division.

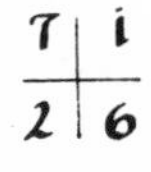

7	1
2	6

The second example

If you have to divide 9065 by 8, arrange your work as follows:

```
9065 | 1
8
```

And first place the divisor, 8, under 9, and say: "8 in 9 gives 1," placing this for the quotient after the bar to the right of the 5. Then multiply this 1 with 8: 1 time 8 is 8, holding this 8 in mind, and subtracting it from 9, saying: "8 from 9," cancelling the 9, saying "1 remainder," and writing the 1 above the 9. Thus is found the first figure. Now proceed further. Arrange your work in this form:

```
1
9065 | 1 1
88
```

F. 27, r.

Now place the divisor, 8, under the 0, and that 1 which stands as a remainder over 9, together with the 0, makes 10. Then we have 8 in 10 gives 1, and we place this 1 after the other, beyond the bar. Then multiply thus: 1 time 8 gives 8, which we cancel. This 8 which came from multiplying we are not able to subtract from 0, but 2, the complement with respect to 10, we write above. Then the 1 of the complementary 10 is taken from the 1 above the 9, thus: 1 from 1 leaves 0, the 1 being then cancelled. Thus is found the second figure. Now to proceed further, arrange the work as follows:

```
12
9065 | 1 1 3
888
```

Now put your divisor, which is 8, under that 6, and, considering that 2 which remains above 0, say: "8 in 26 gives 3," placing this after 11 to the right. Then multiply 8, the divisor, by that 3, thus: 3 times 8 are 24, cancelling the 8 and holding the 24 in mind. Now note here that by the multiplication we produced two figures, namely, 2 for tens and 4 for units, and have at the same time to reduce two figures, the tens and the units; so we reduce the units by units, saying: "4 from 6," cancelling the 6, leaving 2, which 2 we write above the 6. Then reducing tens by tens, saying: "2 from 2," cancelling 2, and there remains 0, and three figures have been found. Now to complete the work, write what has been done in the following form:

```
1 2 2
9 0 6 5 | 1 1 3 3
8 8 8 8
```

F. 27, v.

Now place the divisor, 8, under 5, and then write the 2 which stands above 6, with the 5, making 25, and say: "8 in 25 gives 3 for the quotient," writing it in its place. Now multiply the 8 by 3, saying: "3 times 8 are 24," cancelling the 8 and holding in mind the 24. Now reduce the units by units, saying: "4 from 5 leaves 1," cancelling the 5 and writing the 1 above the 5. Now reduce the tens by tens, saying: "2 from 2 leaves 0," cancelling the 2, and the work is done. Therefore, the result of dividing 9065 by 8 is the quotient 1133 and $\frac{1}{8}$, as is here seen:

```
1 2 2 1
9 0 6 5 | 1 1 3 3
8 8 8 8
```

If you wish to prove that the work is correct, make a cross in which place the excess in the divisor, that is 8, underneath and to the left. Above this place the excess in the quotient, 8. The excess in the remainder, 1, write on the other side and above. Then 8 times 8 are 64, and adding the excess in the remainder, 1, we have 65. Taking the sum of the digits we have 6 and 5 are 11; subtracting 9 we have a remainder of 2, which is the principle excess and is placed under the excess in the remainder. Now see if the excess in the number divided is also 2, and the work is correct.

8	1
8	2

Having explained the galley method of multiplying where the divisor has only one figure, let us attend to an example where the divisor has two figures; thus, if you wish to divide 9875 by 94, arrange the work as follows:

F. 28, r.

```
9 8 7 5 | 1
9 4
```

First put the divisor, 94, under 98. Then say: "9 in 9 is contained 1 time," noticing at the same time that 4 is contained in 8. Hence the quotient is 1, and this is placed after the 5, beyond the bar. Now multiply first 1 by 9, saying: "1 time 9 makes 9," cancelling the 9. Then subtract 9 from 9, cancelling the other 9, the remainder being 0. Then multiply again the quotient 1 by the 4, and say: "1 time 4 is 4." Then subtract 4 from 8, cancelling the 8; 4 remains, and this is written above 8. Thus we have the first quotient. To continue the work, arrange what has been done as follows:

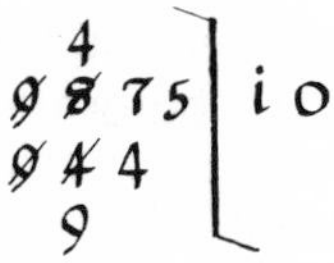

```
 4
9 8 7 5 | 1 0
9 4 4   |
  9     |
```

Now place the divisor, 94, so that 4 is under 7 and 9 under 4, and see if 9 is contained in 4, saying: "9 in 4 gives 0," writing this in the quotient after 1. Then multiply: 0 times 9 makes 0, cancelling the 9 and saying: "0 from 4 leaves 4." Then multiply: 0 times 4 makes 0, cancelling the 4 and subtracting 0 from 7. Thus we have the second quotient. But notice that every time that the quotient is 0 you may cancel the divisor so as to shorten the labor. Now to continue and complete the operation, arrange the work as follows:

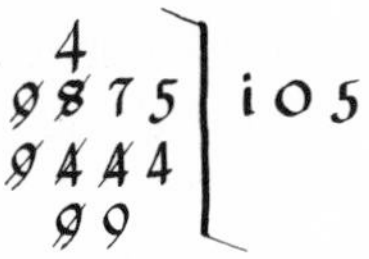

```
 4
9 8 7 5 | 1 0 5
9 4 4 4 |
  9 9   |
```

Then place your divisor, 94, so that the 4 comes under 5, and 9 under the cancelled 4. Note here that just above 9 is 7, and before 7 is 4, which makes 47. Then 9 in 47 gives 5; place this 5 in the quotient after the 0. Now multiply: 5 times 9 are 45, cancelling the 9. Then subtract 5 from 7, cancelling the 7. Write the remainder, 2, above 7. Then reduce 4 by 4, cancelling the 4, and the remainder is 0. Then multiply: 4 times 5 are 20, cancelling 4. Then subtract 0 from 5, leaving 5. Then subtract 2 from 2, cancelling 2, and 0 remains. Thus the work is finished, and the answer is that in

F. 28, v.

dividing 9875 by 94 we get for a quotient 105 and $\frac{5}{94}$, as is seen here below:

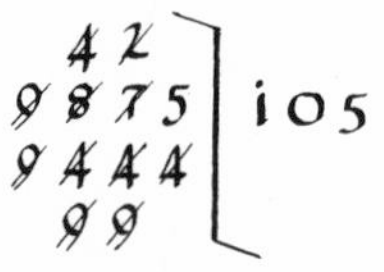

```
  4 2
9 8 7 5 | 105
9 4 4 4 |
  9 9
```

The proof that the work is correct may be effected in the manner above set forth.

Having explained the galley method of dividing when the divisor has two figures, I ask your attention to an example where the divisor has three figures.

If you wish to divide 65284 by 594, arrange the work as follows:

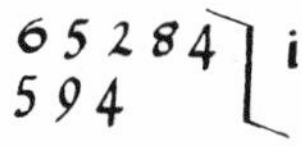

```
6 5 2 8 4 | 1
5 9 4
```

First arrange your divisor in the proper order, as is shown. This being done, see that 5 in 6 gives 1; write 1 in the quotient, beyond the bar. Then multiply each figure of the divisor by this 1, beginning

F. 29, r.

with 5, then taking 9, and then 1, thus: 1 time 5 is 5; cancel 5; then subtract 5 from 6, cancelling 6; then 1 remains, and this 1 is written above 6. Then 1 time 9 is 9, and 9 is cancelled. Then 9 cannot be taken from 5, but 1 is the complement of 9 with respect to 10. Hence cancel the 5, saying: "1 and 5 are 6." Write 6 over 5. Then from the 1 of the complementary 10 subtract 1 which is before 6 and cancel it, since 1 from 1 gives 0. Then multiply: 1 time 4 is 4, and cancel the 4. It is not possible to take 4 from 2, but 6 is the complement of 4 with respect to 10. Hence, cancelling 2 and saying: "6 and 2 are 8," we write 8 above 2; then the 1 of the complementary 10 is taken from 6, saying: "1 from 6," and cancelling 6. There then remains 5, which is written above 6. Thus is found the first quotient. Now to proceed further, arrange the work as follows:

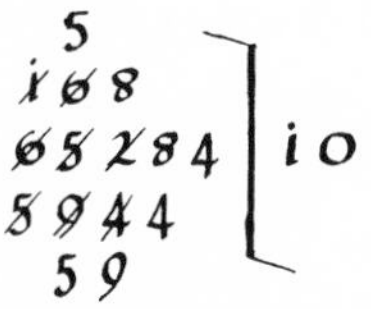

```
  5
 168
 65284 | 10
 5944
  59
```

Then place your divisor in its proper place, 5 under the canceled 9, 9 under the canceled 4, and 4 under 8. Then see if 59 is contained in 58. It is not, since a greater number cannot be contained in a smaller one. Hence we have the quotient 0, which is written after 1, which 0 (according to the statement made above) cancels all the divisor, and is the second quotient. Wishing to continue, write the work in the following form:

F. 29, v.

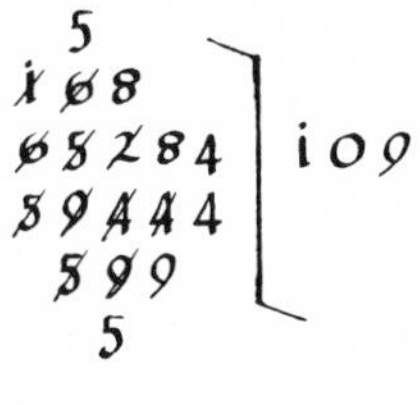

```
  5
 168
 65284 | 109
 59444
  599
   5
```

Now place your divisor in the proper place and proceed as follows: 5 in 58 is 9; place this in the quotient after the 0. Then multiply, saying: "5 times 9 are 45," cancelling the 5. Now take 5 from 8, cancelling the 8, and write the remainder, 3, above 8. Now take 4 from 5, cancelling 5, and write the remainder 1, 1 over 5. Now multiply 9 by 9, giving 81, and cancel the 9 of the divisor. Now take 1 from 8, cancelling 8, and write the remainder, 7, over 8. Since you cannot take 8 from 3, take 2, the complement of 8, and add it to 3, making 5; cancel the 3 and write 5 above it. Then taking the 1 of the complementary 10, subtract it from the 1 above 5, at the same time cancelling 1 from 1, leaving 0. Now multiply, 4 times 9 are 36, and cancel 4. Since you cannot take 6 from 4, add 4, the complement of 6 with respect to 40 [10], thus: 4 and 4 are 8; cancel 4 and write 8 above it. Then take the 4 of the complement with respect to 40 [10] from 7, cancelling 7, and the remainder is 3, and this is written above the canceled 7. The completed work then appears as follows:

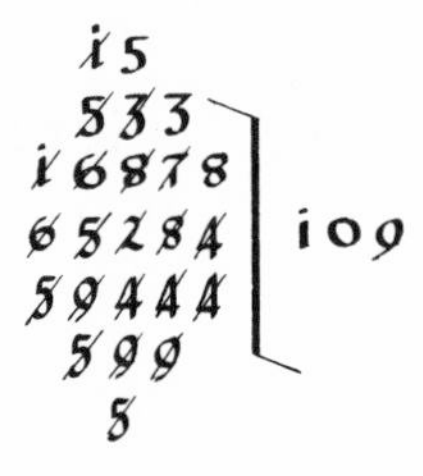

F. 30, r.

This may be proved by casting out 9s in the same way as with the other galley division. What has been given is sufficient for understanding the five operations of the Practica, which are necessary to everyone in mercantile pursuits.

The operations which I have set forth above being understood, it is necessary to take up the method and the rules of using them. That rule you must now study by the rule of the three things. Therefore, that you may have occasion to sharpen your understanding in the four operations above mentioned, addition, subtraction, multiplication, and division, I shall compare them. As a carpenter (wishing to do well in his profession) needs to have his tools very sharp, and to know what tools to use first, and what next to use, etc., to the end that he may have honor from his work, so it is in the work of the Practica. Before you take the rule of the three things it is necessary that you should be very skilled in the operations which have been set forth in addition, subtraction, multiplication, and division, so that you may enter enthusiastically into your work. Furthermore, that the rule of the three things, which is of utmost importance in this art, may be at your command, you must have at hand this tool of the operations, so that you can begin your labors without spoiling your instruments and without failing. Thus will your labors command high praise.

The rule of the three things is this: that you should multiply the thing which you wish to know, by that which is not like it, and divide by the other. And

F. 30, v.

the quotient which arises will be of the nature of the thing which has no term like it. And the divisor will always be dissimilar (in weight, in measure, or in other differences) to the things which we wish to know.

In setting forth this rule, note first that in every case which comes under it there are only two things of different nature, of which one is named twice, by two different numbers, and the other thing is named once, by one number alone. For example:

If 1 pound of saffron is worth 7 pounds of

pizoli, what will 25 pounds of this same saffron be worth? Here are not mentioned together both saffron and money, but the saffron is mentioned twice by two different numbers, by 1 and by 25; and the money is mentioned once, by one number, by 7. So this is not called the rule of the three things because there are three things of different nature, but because one thing is mentioned twice.

Secondly, be it noted that in order to learn to know the three things by their three several names, which are: the thing which has none similar to it, the thing which you wish to know, and the divisor, know that that thing which is mentioned once is the one that is called the thing which has none similar to it, and which is most easily distinguished. And in order that you may be able to recognize the divisor of the thing which you wish to know, note that that thing is always the divisor that manifestly changes or transforms itself into the other thing. Having recognized the names of the two preceding things, the third thing is that which you wish to find. Therefore, in this example, 1 is the divisor, 7 the thing which has none similar to it, and 25

F. 31, r.

is the thing which you wish to know.

Thirdly, it is to be noted that in this particular, as the rule says, the divisor should always be dissimilar to the thing which you wish to know, and indeed it is always so by nature when the two are of one substance, as if I should say in respect to weight or of measure. For even if they be named by various weights, or diverse measures, any two may be reduced to the denomination of the lesser weights or measure. For example, suppose you asked: "If 1 pound of saffron is worth 7 lire of pizoli, what will be the value of 1 ounce?" I say that the divisor, which is now 1 pound, may be reduced to the nature of ounces, saying: "1 time 12 is 12." Then we may say: "If 12 ounces of saffron are worth 7 lire, what will be the value of 1 ounce?" Thus we have arranged it so that two things harmonize in weight, which at first were dissimilar. This being accomplished, we may begin the solution.

And to the end that you may understand the order which I wish to follow in teaching this work, know that I intend to teach it in five ways:

First, with abstract numbers.
Second, with the divisor 1.
Third, with the divisor 100.
Fourth, with the divisor 1000.
Fifth, with any other divisor than those named.

And of each of these five methods I propose only three cases. The first is with respect to numbers entirely integral, except in the fifth method. The second is with respect to numbers partly integral and partly fractional. The third is with respect to numbers that are all fractional.

Now to consider the three cases with respect to the abstract numbers.

First, I ask:

If 8 should become 11, what would 12 become?

F. 31, v.

Secondly, I ask:

If 5 and $\frac{3}{4}$ should become 8 and $\frac{1}{2}$ what would 9 become?

Thirdly, I ask:

If 6 and $\frac{1}{2}$ should become 4 and $\frac{2}{3}$, what would 8 and $\frac{4}{5}$ become? Note that those three cases are called cases of abstract numbers, since here we have no mention of any substance, but only of numbers.

In order to understand the first case, which asks: 'If 8 should become 11 what would 12 become?', note that to give the names to the three terms it is necessary to understand that of these three numbers one is a constant number; and of the other two, one is transformed and the other is to be transformed. Know then that the constant number is the one that has none similar to it, that is 11; and the transformed number is 8, which is the divisor; and 12, which is to be transformed, is the thing which we wish to find. Now, knowing the terms of their names, arrange your work in the following form: first, the divisor; second, the term which has none similar to it; third, the term which we wish to find:

$$\frac{8}{1} \qquad \frac{11}{1} \qquad \frac{12}{1}$$

This being done, know that the numbers which are above the line are numerators, and the numbers which are below the line are denominators; that is to say, that 1 which is under the 8 shows that there are 8 wholes. In the same way the other numbers having 1 beneath are integers, since if they were halves, this would be indicated by 2 beneath, and if they were thirds by 3, etc., as we shall find as we proceed.

F. 32, r.

And before considering the first direction of the rule, it is necessary to do away with the denominators of the term which you intend to use as a divisor. First multiply together the denominator of the term which you intend to use and the denominator of the term which has not one similar to it. Then multiply together the denominator of the term which has not one similar to it, or this number changed, and the numerator of the divisor. If by this the divisor is changed, write directly over the first divisor this number, and what remains is the divisor. Do as follows: multiply the denominator of the divisor by the numerator of the term which has not one like it, and if by this the number is changed write it over the numerator of the term which has not one like it, and this number then represents the term which has not one like it. Having done all this, consider the first operation, which I lay down for you in the rule of 3.

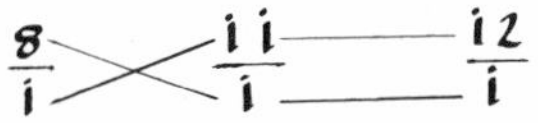

$$\frac{8}{1} \times \frac{11}{1} \text{———} \frac{12}{1}$$

Note that the rule demands of you only two operations: first, that you should multiply; second, that you should divide. Hence, I say first that you should multiply the term which you wish to know, 12, by the term which has not one like it, that is by 11; this being done, arrange your numbers in the form which you have seen in the chapter on multiplication, thus:

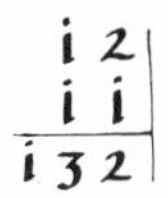

```
 12|
 11|
---
132|
```

The proof of this operation I leave to you.

F. 32, v.

I have now finished this first direction. The

rule next requires the second step, i.e., that you divide this product of the first operation, namely 132, by the other terms, i.e., by 8. This division you should perform by the table method, as I have set forth heretofore, avoiding the galley plan as far as possible. Therefore, arrange your numbers thus:

The divisor	.8.	132	4	The remainder
The quotient		16		

Hence there arises 16 and $\frac{4}{8}$ for the quotient. Answer, therefore, that if 8 becomes 11, 12 will become 16 and $\frac{1}{2}$, and thus is solved the first problem. If you wish to prove that the result is correct, make a cross and cast out 9s. Therefore, multiply first the excess in the term which you wish to know, by that in the term that has none similar to it, thus: 2 times 3 are 6, and this is the principal excess, and is placed below and to the right. Then place the excess in the divisor, 8, underneath and to the left, and the excess of the quotient, 7, above the 8, and the excess of the remainder, 4, above the 6. Then multiply the excess of the divisor by that of the quotient, saying: "7 times 8 are 56," and add the excess of the remainder, 4, giving 60, which (since 0 does not count) makes 6. Hence, the work is right.

The proof:

7	4
8	6

In this way you can prove all cases which come to you. And in the same way, by the method here given, you can solve other similar problems.

F. 33, r.

The second case. If 5 and $\frac{3}{4}$ should become 8 and $\frac{1}{2}$, what would 9 become?

First we have to reduce the integers to the denominations of their fractions, that is 5 and $\frac{3}{4}$ to fourths, saying: "4 times 5 are 20, and 3 are 23, which gives us $\frac{23}{4}$," then 8 and $\frac{1}{2}$ to halves, saying: "2 times 8 are 16, and 1 makes 17, which gives us $\frac{17}{2}$." This being done, put your work in the following form:

If $\frac{23}{4}$ should become $\frac{17}{2}$, what would $\frac{9}{1}$ become?

Then apply the rule, thus:

```
46       68
23 \  /  17          9
-- \/   --          --
4   /\   2 ---------- 1
```

This being done, carry on the work, saying: "1 time 2 is 2; then 2 times 23 are 46," which becomes your divisor. Then 4 times 17 are 68, which becomes the term which has none similar to it. Then follow the rule, and the result will be found, thus:

```
 68 |        1
  9 |       13
----|      254
612 |      612  | 13
           466
            4
```

Therefore, the answer is that if 5 and $\frac{3}{4}$ should become 8 and $\frac{1}{2}$, 9 would become 13 and $\frac{7}{23}$.

The proof:

4	5
1	0

F. 33, v.

The third case. If 6 and $\frac{1}{2}$ should become 4 and $\frac{2}{3}$, what would 8 and $\frac{4}{5}$ become? First reduce the integers to the denominations of their fractions, as you did in the preceding case.

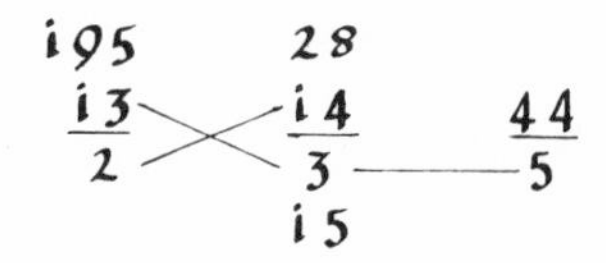

This being done, continue thus: 3 times 5 are 15; write 15 under the denominator of the term that has none similar to it. Then multiply 13 times 15, making 195, which is written above 13, and stands as the divisor. Then 2 times 14 are 28, which is written above 14, and this 28 represents the term which has none similar to it. This being done, carry out the rule as follows:

44
28
1232

6
69
1232 | 6
198

The proof:

6	8
6	8

The quotient is thus seen to be 6 and $\frac{62}{195}$. The answer, therefore, is that if 6 and $\frac{1}{2}$ should become 4 and $\frac{2}{3}$, then 8 and $\frac{4}{5}$ would become 6 and $\frac{62}{195}$. This, then, completes the three cases in abstract numbers.

As to the three cases in which the divisor in the first problem becomes 1.

First question.

If 1 yard of crimson is worth 5 ducats, what will 85 yards be worth?

Second question.

F. 34, r.

If 1 ounce of silver is worth 4 lire and 6 soldi, what will 2 marks and $\frac{1}{2}$ be worth?

Third question.

If 1 pound and $\frac{1}{2}$ of saffron are worth 20 ducats and $\frac{1}{3}$, what will 1 ounce and $\frac{1}{4}$ be worth?

The first case works out as follows:

1 yard of crimson is worth 5 ducats; what will 85 yards be worth? Put your rule in the following shape:

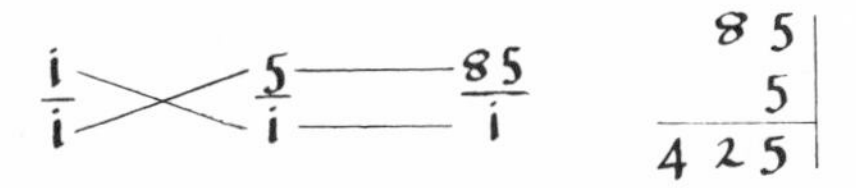

Note that the reasoning is the same here where the divisor is 1, the product or sum of the multiplication being the same as the quotient, since it is not possible to divide by 1. Hence, the quotient turns out to be 425. Therefore, the answer is that 85 yards will be worth 425 ducats.

The proof:

2	0
1	2

The second problem works out as follows:

1 ounce of silver is worth 4 lire and 6 soldi; what will 2 marks and $\frac{1}{2}$ be worth?

Note that the term that has none similar to it is designated by different denominations, so that we need to reduce these to one denomination, namely to the smaller. Multiplying the lire by 20, 4 times 20 are 80, and 6 makes 86, which are 86 soldi. Now arrange your problem again as follows:

1 ounce of silver is worth 86 soldi; what will 2 marks and $\frac{1}{2}$ be worth?

F. 34, v.

Note that since the term which you wish to know is given in integers and fractions, the integers should be reduced to the denominator of the fraction, thus: 2 times 2 are 4, and 1 makes $\frac{5}{2}$.

Now arrange your problem again, as follows:

1 ounce of silver is worth 86 soldi; what will $\frac{5}{2}$ marks be worth? Since the divisor and the term that you wish to find are given in different weights, it is necessary (according to

the statement of the rule of 3) that the larger weight should be reduced to the same unit as the smaller, i.e., marks to the units of ounce. Therefore, since one mark weighs 8 ounces, multiply 5 by 8, saying: "5 times 8 are 40," so that the term you wish to know becomes $\frac{40}{2}$.

This being done arrange your problem as follows:

1 ounce of silver is worth 86 soldi; what will $\frac{40}{2}$ ounces be worth? Now follow your rule, thus:

$$\frac{1}{1} \times \frac{86}{1} \text{———} \frac{40}{2}$$

2 (above), 2 (below)

86		soldi	3440 | 0
40		lire	172(0
3440			86

Now, since the quotient is 1720 soldi, this must be divided by 20. Therefore, note that when you have to divide by 20, cut off the smaller figure, which is the remainder in the soldi, and divide the rest by 2, and the result will be lire. So the answer is that if 1 ounce of silver is worth 4 lire 6 soldi, 2 marks and $\frac{1}{2}$ will be worth 86 lire.

F. 35, r.

The proof:

1	0
2	2

The third problem works out as follows:

If 1 pound and $\frac{1}{2}$ of saffron is worth 2 ducats and $\frac{1}{3}$, what will 1 ounce and $\frac{1}{4}$ be worth?

Note first that the divisor is given in integers and fractions, and it is necessary that the integers be reduced to the same denominator as the fractions, thus: 1 times 2 is 2, and 1 makes 3, which gives $\frac{3}{2}$. This being done, arrange your problem again, thus: $\frac{3}{2}$ pounds of saffron are worth 2 ducats and $\frac{1}{3}$; what will 1 ounce and $\frac{1}{4}$ be worth? Now, since the term which has none similar to it is given in integers and fractions, reduce the integers to the same denominator as the fractions, thus: 2 times 3 are 6, and 1 makes $\frac{7}{3}$. Now arrange your problem again, as follows: $\frac{3}{2}$ pounds of saffron are worth $\frac{7}{3}$ ducats; what will 1 ounce and $\frac{1}{4}$ be worth? Now, since the term like the one which you wish to find is also given in integers and fractions, make all the terms fractional, thus: 1 time 4 is 4, and 1 is $\frac{5}{4}$. Now arrange your problem again, thus: $\frac{3}{2}$ pound of saffron is worth $\frac{7}{3}$ ducats; what will $\frac{5}{4}$ ounce be worth? Now, since the divisor and the term like the one you wish to find

F. 35, v.

are given in different weights, it is necessary

that the greater weight should be reduced to the smaller denomination, that is, the number of the divisor, which is pounds, must be reduced to ounces. Then, since 12 ounces makes 1 pound, multiply this 3, the numerator, of the divisor, by 12, saying: "3 times 12 are 36." Then arrange again your problem as follows: $\frac{36}{2}$ ounces of saffron are worth $\frac{7}{3}$ ducats; what will $\frac{5}{4}$ ounces be worth? Now proceed to apply your rule, remembering the directions regarding the rule of 3.

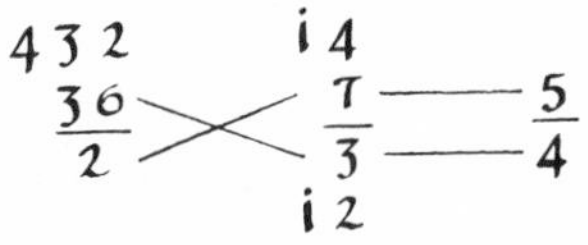

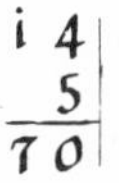

```
432       14
 36 \/    7 ——— 5
 2  /\    3 ——— 4
          12
```

```
 14|
  5|
 70|
```

Note here that since 70 is less than the divisor, it is not possible to perform the division. This shows that the price of 1 and $\frac{1}{4}$ ounce cannot amount to a ducat. On this account, multiply 70 by 24 to reduce to golden grossi; and then divide by 432, and the quotient will be grossi.

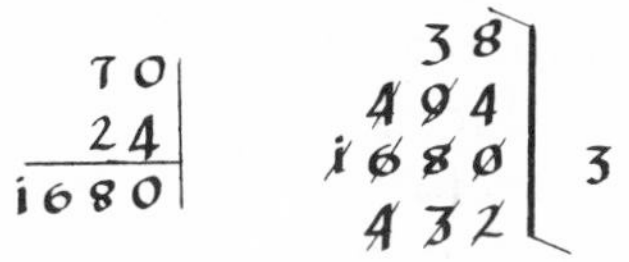

```
  70|       38
  24|      494
1680|     1680 | 3
           432
```

F. 36, r.

Therefore, the quotient is 3 grossi. And since you see that the preceding galley division gives for a remainder 384,

note that it is necessary to multiply this by 32, and then divide by 432, the quotient then being pizoli.

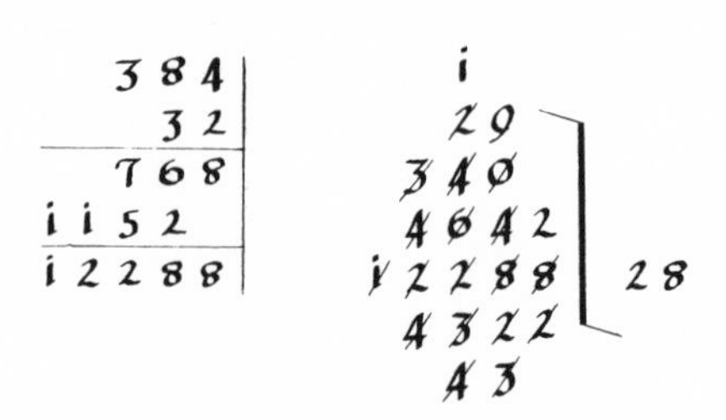

Thus we have 28 pizoli, and the problem is solved. Therefore, the answer is that if 1 pound and $\frac{1}{2}$ of saffron is worth 2 ducats and $\frac{1}{3}$, 1 ounce and $\frac{1}{4}$ will be worth 3 grossi, 28 pizoli and $\frac{4}{9}$. To prove this result you should notice that it is necessary to multiply the term that has none like it, 14, first by the excess in the radix 24, saying: "5 times 6 are 30, the excess in which is 3." Then multiply this 3 by the excess in the radix 32, saying: "3 times 5 are 15, the excess in which is 6." Then multiply this 6, which is left as the excess of the term like that which you wish to find. This change of proof is necessary for the reason that the money found in the quotient is of less denomination than that mentioned in the problem, since ducats are given and here we have pizoli. Therefore, note carefully that this method is to be followed in all similar cases.

The proof:

7	3
0	3

Thus is completed the discussion in which the divisor in the first problem is 1.

Hence by these methods (if you have been studious) may be understood all problems of similar or varied kind. Note that from now on I do not intend to explain anything that I have already set forth, so be studious and attentive, to the end that you may clearly understand what has been passed over.

F. 36, v.

As to the third kind of problem, in which the divisor in the first case is 100.

First question.

If 100 pounds of sugar are worth 32 ducats, what will 9812 pounds be worth?

Second question.

If 100 pounds of ginger are worth 16 ducats, 18 grossi, 14 pizoli and $\frac{1}{2}$, what will 8564 pounds and $\frac{1}{3}$ be worth?

Third question.

If 100 pounds and $\frac{1}{4}$ of silk are worth 42 ducats, 7 grossi and $\frac{1}{5}$, what are 9816 pounds, 3 ounces and $\frac{1}{6}$ worth?

The first case works out as follows:

If 100 pounds of sugar are worth 32 ducats, what will 9812 pounds be worth? Arrange your rule as follows:

$$\frac{100}{1} \times \frac{32}{1} — \frac{9812}{1}$$

```
            9812|
              32|
          ------|
           19624|
          29436 |
ducats    ------|
          3139(84|
```

Understand here that whenever your divisor has to the right-hand any 0s, as many 0s as you have, so many figures must you cut off from the number to be divided, and the rest of the number you should divide by

F. 37, r.

the significant figure or figures which are in the divisor. So, if your divisor is 100, as here, cut off the last two figures, and what is left is the quotient. And when your divisor is 200, cut off two figures and divide what is left by 2; if 300, by 3; if 1000, cut off three figures; if 2000, cut off these figures and divide the rest by 2; if 3000, by 3. Do the same with other numbers of this class, known as articles, when you find them. In the present problem, cut off two figures, and what is left besides the part cut off represents the quotient, 3139, which is ducats, and the two figures cut off, namely, 84, are the remaining ducats which you must reduce to grossi, multiplying these by 24, and then dividing by the divisor, 100. Therefore, multiply:

```
           8 4 |
           2 4 |
grossi  20(1 6 |
```

This 16 is the remainder from the grossi, which you must multiply by 32 and then divide by 100 to reduce to pizoli. Therefore, multiply:

```
          3 2 |
          1 6 |
pizoli   5(1 2 |
```

This 12 is the remainder. Thus, the work is finished. Therefore, the answer is that if 100 pounds of sugar are worth 32 ducats, 9812 pounds will be worth 3139 ducats, 2 [20] grossi, 5 pizoli and $\frac{3}{25}$.

```
The proof:   0 | 3      7 | 3
             --+--      --+--
             2 | 3      1 | 1
```

F. 37, v.

The second case works out as follows:

If 100 pounds of ginger are worth 16 ducats, 18 grossi, 14 pizoli and $\frac{1}{2}$, what will 8564 pounds and $\frac{1}{3}$ be worth? Arrange your rule as follows:

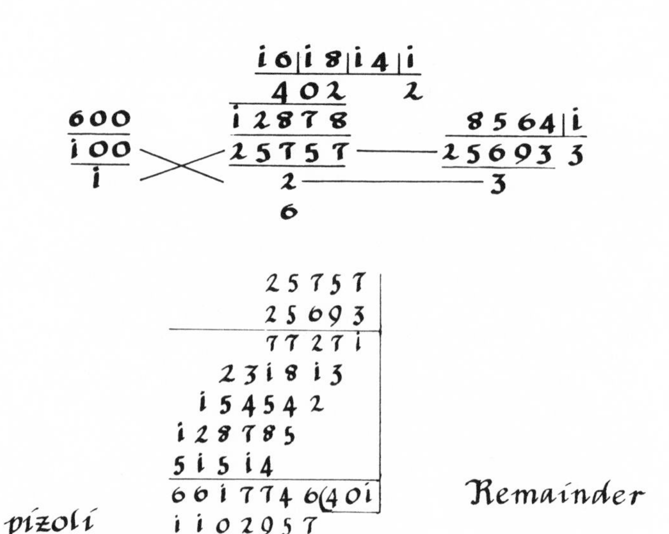

Note that when you come to a certain amount in pizoli, which can be expressed in units of a higher denomination, you should reduce the pizoli to as high a denomination as possible. Take, for instance, this number 1102957, the number of pizoli in this case, and divide it by 32 to reduce to grossi, thus:

1102957 13 pizoli
grossi 34467

Then you should take the 34467, which now represents grossi, and divide it by 24, to get ducats, thus:

34467 3 grossi
ducats 1436

F. 38, r. Thus your problem is solved, and the

answer is that if 100 pounds of ginger are worth 16 ducats, 18 grossi, 14 pizoli and $\frac{1}{2}$, 8564 pounds and $\frac{1}{3}$ are worth 1439 [1436] ducats, 3 grossi, 13 pizoli and $\frac{401}{600}$.

The proof:

7	5
6	2

The third case works out as follows: If 100 pounds and $\frac{1}{4}$ of silk are worth 42 ducats, 7 grossi and $\frac{1}{5}$, what are 9816 pounds, 3 ounces and $\frac{1}{6}$ worth? Arrange your rule as follows:

```
                  20304
12030             42|7|1          9816|3|1
                        5                 6
        100 1     1015            117795
            4
        401       5076            706771
         4          5                  6
                   30
```

```
      706771
       20304
   282708 4
  2120313 0
 14135420
 14350278384
```

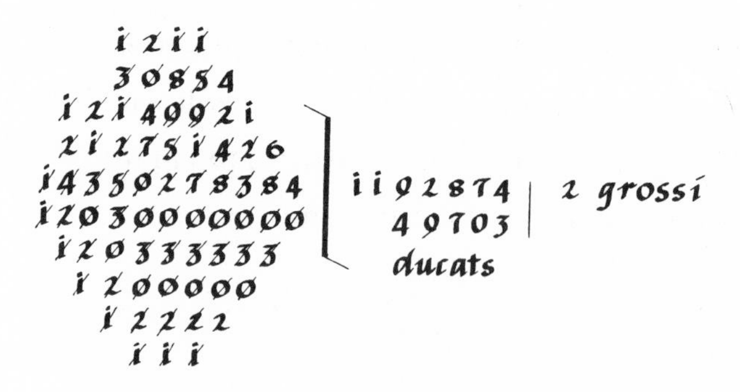

F. 38, v.

4164
32
8328
12492
133248

1291
133248 | 11 pizoli
120300
1203

So the problem is solved, and the answer is that if 100 pounds and $\frac{1}{4}$ of silk are worth 42 ducats, 7 grossi and $\frac{1}{5}$, 9816 pounds, 3 ounces and $\frac{1}{6}$ are worth 703 [49703] ducats, 2 grossi, 11 pizoli and $\frac{153}{2005}$.

The proof:

0	0
6	0

[The author has forgotten to convert from ounces to pounds. The correct answer is 4141 ducats, 22 grossi, 5 pizoli and $\frac{1}{3}$.]

As to the fourth style of problem promised, in which the divisor in the first case is 1000.

First question.

If 1000 pounds of French wool are worth 120 ducats, what will 10292 pounds be worth?

Second question.

If 1000 pounds of pepper are worth 80 ducats, 16 grossi and $\frac{1}{4}$, what will 9917 pounds and $\frac{1}{2}$ be worth?

Third question.

If 1000 pounds and $\frac{1}{5}$ of cinnamon are worth 130 ducats and $\frac{1}{4}$, what will 14616 pounds, 9 ounces, 5 sazi and $\frac{1}{3}$ be worth?

The first case works out as follows:

If 1000 pounds of French wool are worth 120 ducats, what will 10292 pounds be worth? Arrange the rule as follows:

$$\frac{1000}{1} \quad \frac{120}{1} \quad \frac{10292}{1}$$

F. 39, r.

```
           10292
             120
         -------
          205840
         10292
         -------
ducats   1235(040

           40
           24
         ----
grossi   0(960

            960
             32
         ------
           1920
          2880
         ------
pizoli   30(720
```

Thus the problem is solved, and the answer is that if 1000 pounds of French wool are worth 120 ducats, 10292 pounds will be worth 1235 ducats, 0 grossi, 30 pizoli and $\frac{18}{25}$.

The proof:

0	0
1	0

The second case works out as follows:

If 1000 pounds of pepper are worth 80 ducats, 16 grossi and $\frac{1}{4}$, what will 9917 pounds and $\frac{1}{2}$ be worth? Arrange the rule as follows:

```
                    1
 8000        80|16|4         1
 1000        1936        9917|2
 ----        7745 ------ 19835
   1           4  ------   2
               8
```

F. 39, v.

```
      19835
       7745
     ------
      99175
     79340
   138845
  138845
  153622(6075

                                          6075
                                            32
            19202 |   2 grossi           12150
 ducats       800 |                     18225
                                        194(2400
                         pizoli           24
```

Thus the problem is solved and the answer is that if 1000 pounds of pepper are worth 80 ducats, 16 grossi, and $\frac{1}{4}$, 9917 pounds and $\frac{1}{2}$ will be worth 800 ducats, 2 grossi, 24 pizoli and $\frac{3}{10}$.

The proof:

4	6
8	2

The third case works out as follows:

If 1000 pounds and $\frac{1}{5}$ of cinnamon are worth 130 ducats and $\frac{1}{4}$, what are 14616 pounds, 9 ounces, 5 sazi and $\frac{1}{3}$ worth? Arrange the rule as follows:

```
43200864   [4320864]
 7200144    [720144]
  600012     [60012]

                                 14616|9|5|1
                                            3
                   2605
   1000 1          130 1          17540 1
        5              4          1052411
   5001            521 ---------- 3157234
   5               4 ------------ 3
                   12
```

F. 40, r.

```
    3157234
       2605
   15786170
 189434040
 6314468
 8224594570
```

```
        19
        2090
        149523
     233501657
    4903730998
    8224594570
   43208644444
    43208666
      432088
        4320
```

1903 ducats

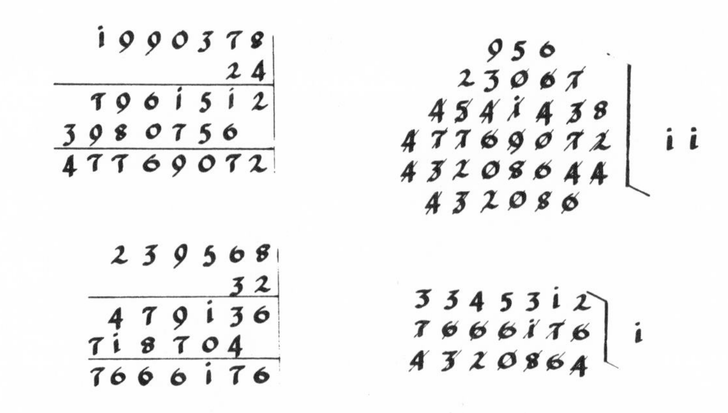

So the problem is solved. Therefore, if 1000 pounds and $\frac{1}{5}$ of cinnamon are worth 130 ducats and $\frac{1}{4}$, 14616 pounds, 9 ounces, 5 sazi and $\frac{1}{3}$ are worth 1903 ducats, 11 grossi, 1 pizoli and $\frac{3345312}{4320864}$.

F. 40, v.

As to the fifth kind of problem promised above, there are three cases, by which you must know that one case proves the other, namely when you make of what is demanded (mentioning its price) the divisor, and of the divisor the thing demanded, from what is demanded comes its first price, both cases being proved by one another.

And I wish (to approach the matter more closely) to take up the three cases of the fourth kind and reconsider them.

The first case is this: If 1000 pounds of French wool are worth 120 ducats, what will 10292 pounds be worth? The answer is that it is worth 1235 ducats, 0 grossi, 30 pizoli and

$\frac{720}{1000}$. Having considered this, state the first case thus: If 10292 pounds of French wool are worth 1235 ducats, 0 grossi, 30 pizoli and $\frac{720}{1000}$, what will 1000 pounds be worth?

The second case if this: If 1000 pounds of pepper are worth 80 ducats, 16 grossi and $\frac{1}{4}$, what will 9917 pounds and $\frac{1}{2}$ be worth? The answer is that they are worth 800 ducats, 2 grossi, 24 pizoli and $\frac{3}{10}$. Having considered this, state the second case thus: If 9917 pounds and $\frac{1}{2}$ of pepper are worth 800 ducats, 2 grossi, 24 pizoli and $\frac{3}{10}$, what will 1000 pounds be worth?

F. 41, r.

The third case is this: If 1000 pounds and $\frac{1}{5}$ of cinnamon are worth 130 ducats and $\frac{1}{4}$, what will 14616 pounds, 9 ounces, 5 sazi and $\frac{1}{3}$ be worth? The answer is that they are worth 1903 ducats, 11 grossi, 1 pizoli and $\frac{3345312}{4320864}$. Having considered this, state the third case thus: If 14616 pounds, 9 ounces, 5 sazi and $\frac{1}{3}$ [of cinnamon] are worth 1903 ducats, 11 grossi, 1 pizoli and $\frac{3345312}{4320864}$; what are 1000 $\frac{1}{5}$ pounds worth?

The first case works out as follows: 10292 pounds of French woolen are worth 1235 ducats, 0 grossi, 30 pizoli and $\frac{720}{1000}$; what are

1000 pounds worth? Arrange the rule as follows:

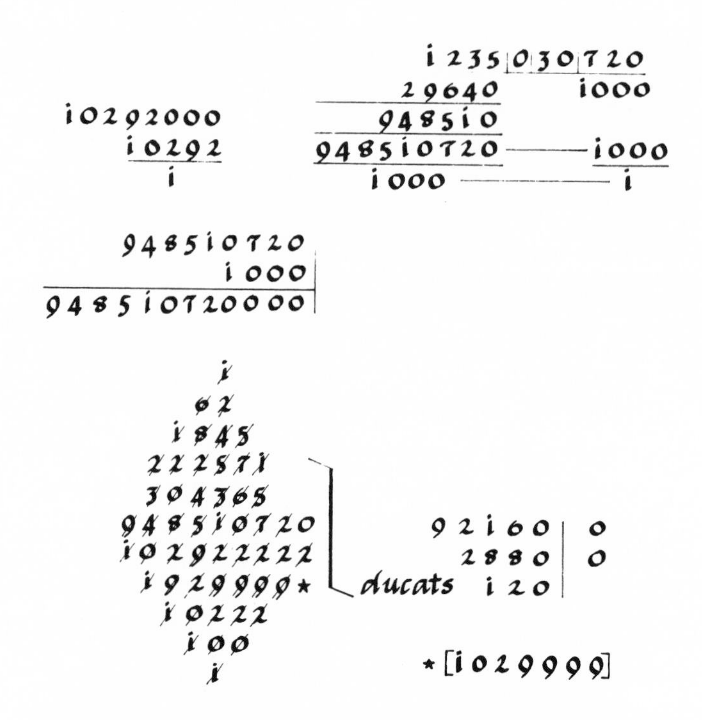

F. 41, v.

Thus the problem is solved, and the result turns out as promised, if 10292 pounds of French wool are worth 1235 ducats, 0 grossi, 30 pizoli and $\frac{720}{1000}$, 1000 pounds will be worth 120 ducats. And this is its price, and so you can understand the problem and see that the result is right.

Therefore, you can prove in this way all other problems, turning them about as you see in this case.

The second case works out as follows:

If 9917 pounds and $\frac{1}{2}$ of pepper are worth 800 ducats, 2 grossi, 24 pizoli and $\frac{3}{10}$, what are 1000 pounds worth?

Arrange the rule as follows:

```
                     12289766
                                      3
198350                   800|2|24|10
                        19202
  9917 1                614488
       2
19835                  6144883 ———— 1000
    2                    10    ———— 1

              12289766|
                  1000|
          12289766000|
```

F. 42, r.

```
        8
       11
      102
     29952
     30090
     498243
    6801411
  12289766000        61960 | 8  pizoli
   1983555555         1936 | 16 grossi
    1983333      ducats 80 |
     19888
      199
       1
```

Thus the problem is solved, and it comes out as promised, i.e., at the price mentioned; that if 9917 pounds and $\frac{1}{2}$ of pepper are worth

800 ducats, 2 grossi, 24 pizoli and $\frac{3}{10}$, then 1000 pounds will be worth 80 ducats, 16 grossi, 8 pizoli, which are $\frac{1}{4}$ of one grosso. Thus these two problems are correct.

The third case works out as follows:

If 14616 pounds, 9 ounces, 5 sazi and $\frac{1}{3}$ [of cinnamon] are worth 1903 ducats, 11 grossi, 1 pizoli and $\frac{3345312}{4320864}$, what will 1000 pounds and $\frac{1}{5}$ be worth? Arrange the work as follows:

F. 42, v.

```
                    3345312
1903 | 11 | 1 |     4320864
  45683
1461857
```

```
      4320864
      1461857
    30246048
   216043202
  345669121
  43208643
 259251845
1728345 94
43208643
       3
6316488629760
```

Multiply by the term that has none similar to it.

[172834504]

```
6820989365 0880
                                                            360072
  14616|9|5|1                                               60012
 175401     3                                               1001
1052411        18949465889280                               5
3157234        6316488629760 ---------------------------    5001
   5 [3]         4320864                                    3 [5]
```

Multiplied by
to make the divisor.

```
        21604320
         3157234
        86417280
       64812960
      43208640
     151230240
    108021600
    21604320
   64812960
   68209893650880
```

(For the numerical work, see the illustration.)

F. 43, r.

```
          18949465889280
                  360072
         37898931778560
       132646261224960
  11369679533568000
  56848397667840
  6823172081684828160
```

```
    1    1
    13   21181
    346  49707
    1824128981  11
   2293265063047
 6823172081684828160
 6820989365088000000      100032 | 0
  682098936508888888        3126 | 6
   68209893650888 8       ducats 130
    6820989365000
     682098936 55
      6820989 36
```

Thus the problem is solved, and it comes out as promised, i.e., at the price mentioned; that if 14616 pounds, 9 ounces, 5 sazi and $\frac{1}{3}$ [of cinnamon] are worth 1903 ducats, 11 grossi, 1 pizoli and $\frac{3345312}{4320864}$, 1000 pounds and $\frac{1}{5}$ will be worth 130 ducats, 6 grossi, which are a fourth of one ducat. Thus that problem and this one are both correct.

Notice that when you have to solve any problem of importance, and where you have any doubts, there is no better proof than to reverse the given problem in the way shown in the three preceding cases.

Hence, by this and by the other problems already stated, which number thirteen in all, you can understand

F. 43, v.

sufficiently the method of solving all such cases as will come to you in your business.

And this sufficeth as to the things already promised by me.

Notwithstanding the things already set forth, I wish now (to satisfy the more completely your longing and endeavor) to teach you five other styles of problem, and to this I invite your attention.

And first I wish to show you certain problems in the rule of three things, in which (notwithstanding the aforesaid rule of three things) if you divide the product of the multiplication of the other terms by the term which is interchanged with the one which has none similar to it, the resulting quotient will be of the same nature as the term which you seek.

Second : I shall teach you to discount tare and tret.
Third : Problems in partnership.
Fourth : Problems in barter.
Fifth : Problems in alloys of coins.

And concerning each of these topics I shall show you only three problems. As to the three problems under the first topic:

First problem.

When a bushel of wheat is worth 8 lire, the bakers make a loaf of bread weighing 6 ounces; required is the number of ounces in the weight of a loaf when it is worth 5 lire a bushel.

Second problem.

I have 16 florins worth 4 lire and 12 soldi and $\frac{1}{2}$ each, which I wish to change into ducats worth 5 lire, and 14 soldi and $\frac{1}{3}$ each. How many shall I have?

Third problem.

F. 44, r.

I have 9 yards and $\frac{2}{3}$ of cloth 2 yards and $\frac{3}{4}$ wide, which I wish to make into a garment. I wish to line this with cloth 1 yard and $\frac{1}{8}$ wide; how much do I need?

As to the first problem, arrange the rule as follows:

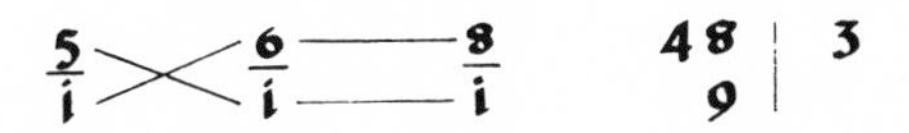

Thus the problem is solved, and the answer is that the bakers ought to make loaves weighing

9 ounces and $\frac{3}{5}$, when grain is worth 5 lire a bushel.

The second problem arranges itself as follows:

I have 16 Rhenish florins worth 4 lire, 12 soldi, and $\frac{1}{2}$ each, and I wish to change them into ducats at 5 lire, 14 soldi, and $\frac{1}{3}$ each; required is how many are due me. Arrange the rule as follows:

F. 44, v.

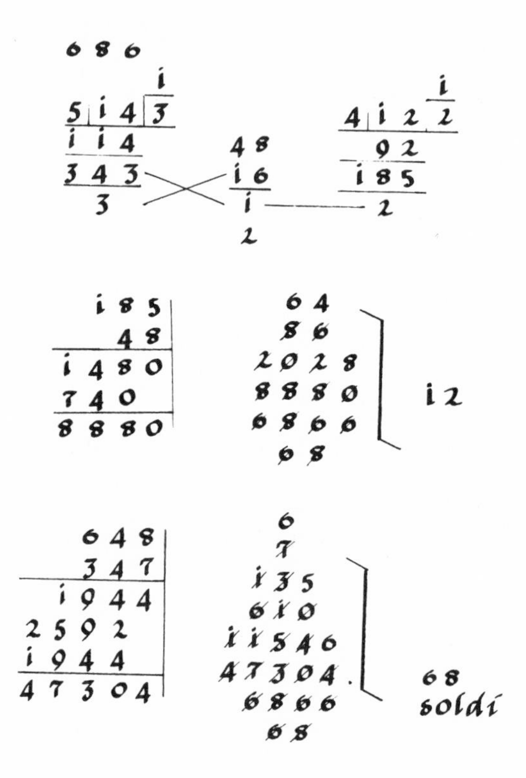

Thus the problem is solved, and the answer is that I should receive 12 ducats, 3 lire, 8 soldi and $\frac{328}{343}$. [*The author made a place-value error in the third partial product; thus the remainder of the solution and the resulting answer are incorrect. The correct answer is 12 ducats, 5 lire, 8 soldi.]

The third problem arranges itself as follows:

I have bought 9 yards and $\frac{2}{3}$ of cloth 2 yards and $\frac{3}{4}$ wide, wishing to make a garment. I wish to line it with cloth 1 yard and $\frac{1}{8}$ wide. Required is the amount of lining needed. Arrange the rule as follows:

```
      1         2        3
    1 —       9 —      2 —
      8         3        4
  1 0 8     2 3 2
    9        2 9       1 1
    —   ✕    —    ———  — 
    8        3         4
            1 2
```

```
    2 3 2            3
      1 1          6 7 4
  -------        1 5 5 2      16
    2 3 2        1 8 8 0      yards
  2 3 2            1 0
* 1 5 5 2
```

Thus the work is completed, and the answer is that I shall need 16 yards and $\frac{17}{54}$ of this cloth.

[*The author's solution is wrong. It should be 2552 ÷ 108, giving 23 yards and $\frac{17}{27}$.]

F. 45, r.

As to the three problems of the second topic, i.e., tare and tret:

First problem.

If a hundredweight of yarn is worth 8 [18] ducats, what is the value of the tare, at 4 pounds per hundred, on 4562 pounds?

Second problem.

If a hundredweight of cotton is worth 36 ducats, 10 grossi, 10 pizoli, what is the amount to be deducted for 8348 pounds, tare being 6 pounds per hundred, and tret 2 ducats per hundredweight?

Third problem.

If a hundredweight of wool is worth 19 ducats, 14 grossi and $\frac{1}{2}$, what should be deducted for 9968 [4562] pounds, tare being 3 per cent, and tret 2 and $\frac{3}{4}$ [per cent]?

The first problem arranges itself as follows:

If a hundredweight of yarn is worth 18 ducats, what is the value of the tare, at 4 pounds per hundred, on 4562 pounds?

Note carefully that this tare is deducted from the 4562 pounds, whence the rule of 3 applies, which says: If a hundredweight is reduced 4, what will 4562 pounds be reduced? Therefore seek the tare which has to be allowed, arranging the rule as follows:

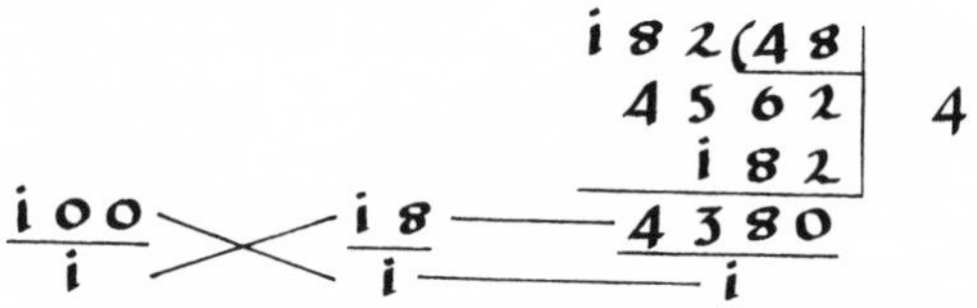

```
                        182(48 |
                         4562  |   4
                          182  |
100 \/ 18 ------------   4380
 1  /\  1 ------------      1
```

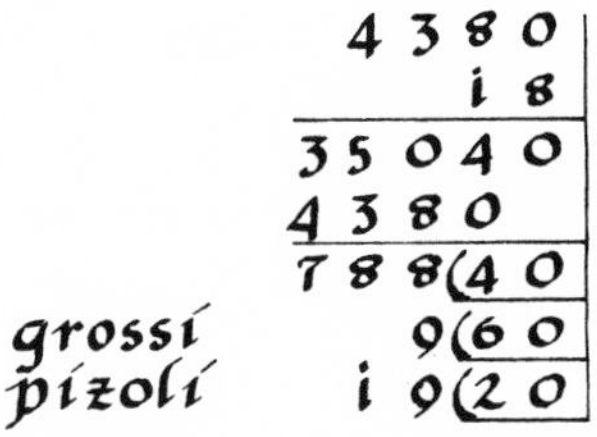

	4380
	18
	35040
	4380
	788(40
grossi	9(60
pizoli	19(20

F. 45, v.

Thus the problem is solved, and the answer is that if a hundredweight of yarn is worth 18 ducats, 4562 pounds having a deduction of 4 pounds per hundred for tare, the value of the tare is 788 ducats, 9 grossi, 19 pizoli.

Note here that when you have to make a reduction for tare, you should see if your remainder exceeds half of the divisor, since if it does you should add 1 to the tare. But if it does not exceed half of the divisor, as here, where the remainder is only 48 on 100, this remainder is not considered, as you will understand if you consider the method that I have set forth.

The second problem arranges itself as follows:

The hundredweight of cotton is worth 36 ducats, 10 grossi, 10 pizoli; what is the amount to be deducted for 8348 pounds, tare being 6 pounds per hundred, and tret 2 ducats per hundred [weight]?

Deduct the tare, and arrange the work as follows:

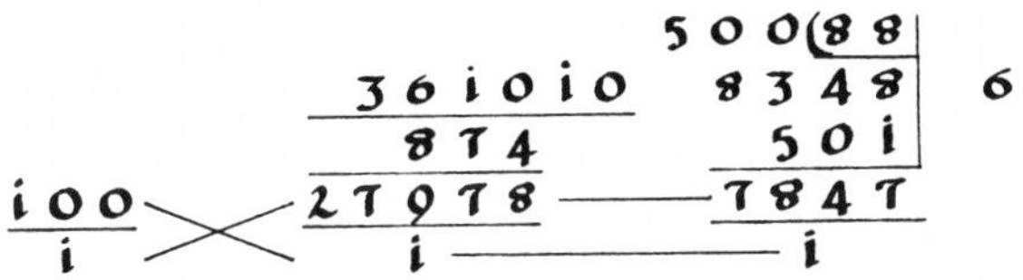

			500(88	
	361010		8348	6
	874		501	
100	27978		7847	
1	1		1	

Note that the tret is deducted from the price of the merchandise thus:

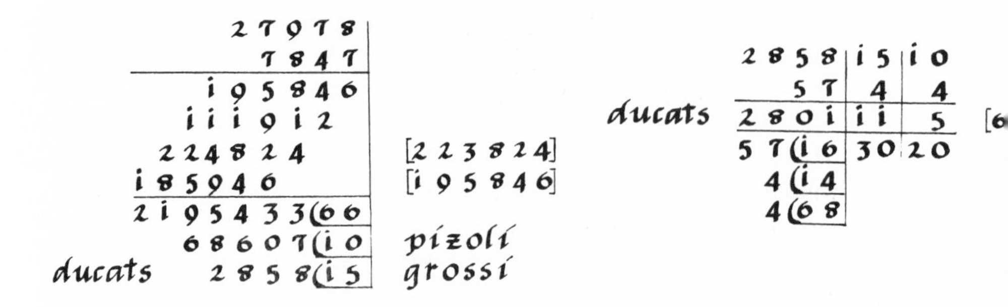

F. 46, r.

The answer is that if a hundredweight of cotton is worth 36 ducats, 10 grossi, 10 pizoli, there should be deducted on 8348 pounds, the tare being 6 pounds per hundred and the tret 2 ducats per hundred, 2801 ducats, 11 grossi, 5 pizoli.

The third problem arranges itself as follows:

The hundredweight of wool is worth 19 ducats, 14 grossi and $\frac{1}{2}$; what should be deducted for 4562 pounds, tare being 3 per cent, and tret 2 and $\frac{3}{4}$ [per cent]?

```
                              136(86
                 19|14|1      45623
 200             470      2     137
 100             941 ———————   4425
   1               2 ———————      1
```

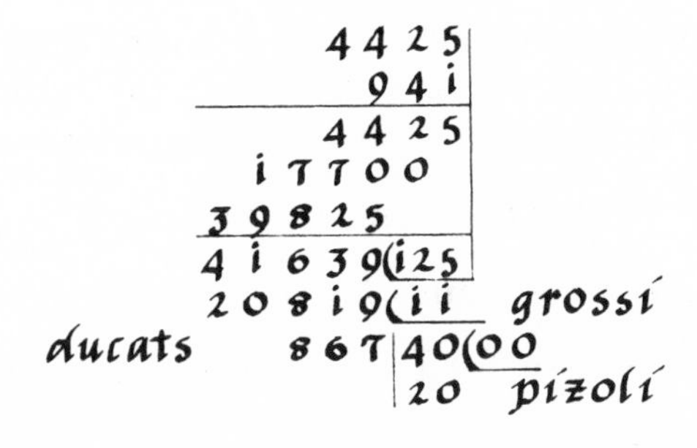

Note that to deduct tret at 2 and $\frac{3}{4}$, or a similar fraction, you should multiply the number to be discounted by the integer, which is here 2, and then divide this same number by 4, which is the denominator of the fraction, and place this quotient under the product of the multiplication by 2, and then multiply this quotient by the numerator of the fraction, which is 3, less 1, which is 2, i.e., by 2, and then add these three numbers together. These being added, you should divide by your principal divisor of the tret as is set forth in the following:

F. 46, v.

	867	11	20
	23	20	17
ducats	843	15	3
	1734	23	8
	216	20	29
	433	17	26
	23(85	13	31
	20(53		
	17(27		

Thus the problem is solved, and the answer is that which you find to the right of the word "ducats," that is, if a hundredweight of wool is worth 19 ducats, 14 grossi, and $\frac{1}{2}$, which has a discount of 3 per cent for tare and 2 and $\frac{3}{4}$ per cent for tret, 4562 pounds will be worth 843 ducats, 15 grossi, 3 pizoli.

Now as to the third class of problems, in partnership:

First problem.

Three merchants have invested their money in a partnership, whom to make the problem clearer I will mention by name. The first was called Piero, the second Polo, and the third Zuanne. Piero put in 112 ducats, Polo 200 ducats, and Zuanne 142 ducats. At the end of a certain period they found that they had gained 563 ducats. Required is to know how much falls to each man so that no one shall be cheated.

Second problem.

Two merchants, Sebastiano and Jacomo, have invested their money for gain in a partnership. Sebastiano put in 350 ducats on the first day of January, 1472, and Jacomo 500 ducats, 14 grossi on the first day of July, 1472; and on the first day of January, 1474 they found that they had gained 622 ducats. Required is the share of each.

F. 47, r.

Third problem.

Three men, Tomasso, Domenego, and Nicolo, entered into partnership. Tomaso put in 760 ducats on the first day of January, 1472, and on the first day of April took out 200 ducats. Domenego put in 616 ducats on the first day of February, 1472, and on the first day of June took out 96 ducats. Nicolo put in 892 ducats on the first day of February, 1472, and on the first day of March took out 252 ducats. And on the first day of January, 1475, they found that they had gained 3168 ducats, 13 grossi and $\frac{1}{2}$. Required is the share of each, so that no one shall be cheated.

The first problem arranges itself as follows:

Three merchants, viz., Piero, Polo, and Zuanne, went into partnership. Piero put in 112 ducats, Polo 200 ducats, and Zuanne 142 ducats. They found that they had gained 53 [563] ducats. Required is the share of each.

In this and all problems of partnership you set down all of the shares one after the other, and find their sum, which becomes your divisor, thus:

Piero put in	112	ducats
Polo put in	200	ducats
Zuanne put in	142	ducats
The sum	454	Divisor

Then take the case of Piero, saying: "If 454 ducats gain me 563 ducats, how much should

F. 47, v.

112 ducats gain me?" Now you know your divisor. Setting forth the rule in form, you know what should be done according to its directions.

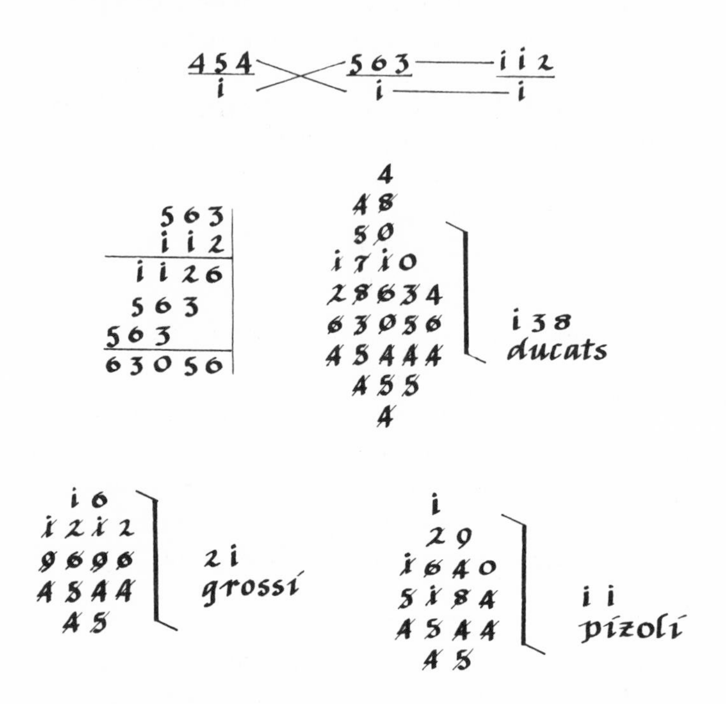

Thus the problem is solved, and the answer is that there falls to Piero as profit 138 ducats, 21 grossi, 11 pizoli and $\frac{190}{454}$.

Then take the case of Polo, saying: "If 454 ducats gain me 563 ducats, how much should

200 ducats gain me?" Arrange the rule as follows:

F. 48, r.

454 — 563 — 200
1 — 1 — 1

34
256
31848
112600
45444
455
4

248
ducats

563
200
112000

24
148
2002
6144
4544
45

13
pizoli

[remainder from division,
8, converted into grossi,
$8 \times 24 = 192$;
$192 \div 454 = 0$ grossi]

192
grossi 0

Thus the work is finished, and the answer is that Polo's share of profit is 248 ducats, 0 grossi, 13 pizoli and $\frac{242}{454}$.

Then take the case of Zuanne, saying: "If 454 ducats gain me 563 ducats, how much should 142 ducats gain?" Arrange the rule as follows:

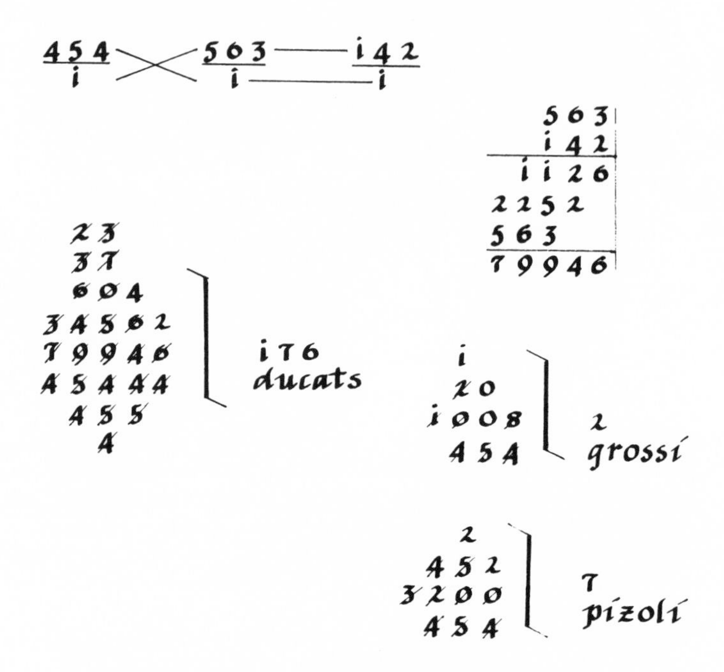

Thus the work is finished, and the answer is that Zuanne's share of profit is 176 ducats, 2 grossi, 7 pizoli and $\frac{22}{454}$.

To prove the cases of all three, that no one has been

F. 48, v.

cheated, take the sum of the profits of all three. Since these amount to exactly 563 ducats, which is the total given, no one has been cheated.

Piero gains
ducats 138 *grossi* 21 *pizoli* 11 $\frac{190}{454}$

Polo gains
ducats 248 *grossi* 0 *pizoli* 13 $\frac{242}{454}$

Zuanne gains
ducats 176 *grossi* 2 *pizoli* 7 $\frac{22}{454}$

Sum of the gains
ducats 563 *grossi* 0 *pizoli* 0 $\frac{0}{454}$

Thus this problem in partnership is shown to be correct.

The second problem arranges itself as follows:

Two merchants, Sebastiano and Jacomo, enter into partnership. Sebastiano put in 350 ducats on the first day of January, 1472; Jacomo put in 500 ducats, 14 grossi on the 1st day of July, 1472. On the 1st day of January 1474 they find that they have gained 622 ducats. Required is the share of each.

Sebastiano put in 350 ducats.

Jacomo put in 500 ducats, 14 grossi.

Note carefully that the two shares should be reduced to grossi, since we should always allow for the difference in denominations. Therefore, write down the shares here, and reduce to grossi.

350	
500	14
8400	
12014	

And since Sebastiano has had his share in 6 months longer than Jacomo, we must multiply each share by the length of its time, thus:

F. 49, r.

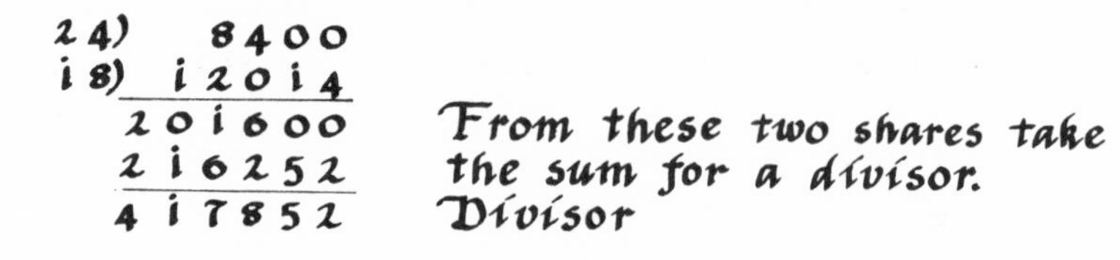

Now apply the rule for Sebastiano, as follows.

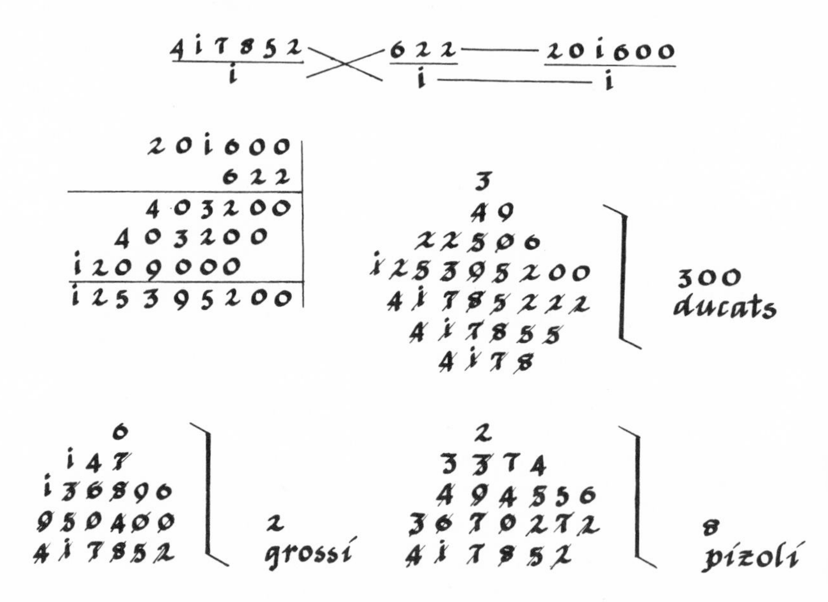

Thus the problem is solved, and the answer is that Sebastiano's share of profit is 300 ducats, 2 grossi, 8 pizoli and $\frac{327456}{417852}$. Now apply the rule for Jacomo, as follows:

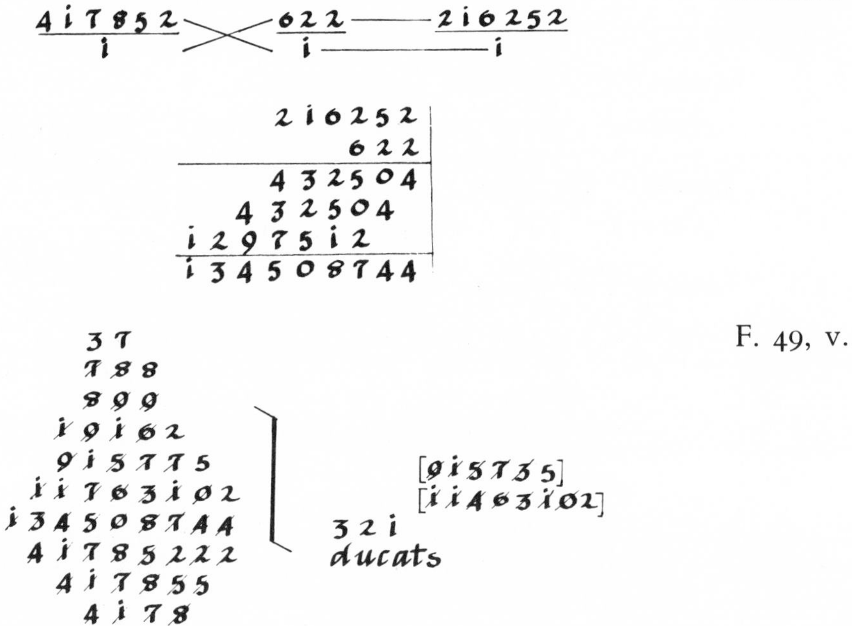

F. 49, v.

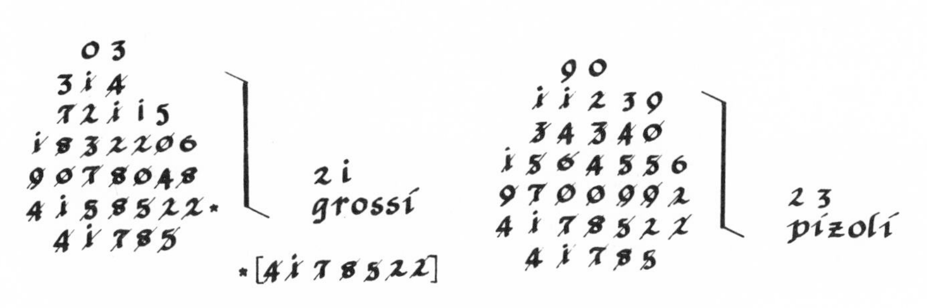

Thus the problem is solved, and the answer is that to Jacomo comes the profit of

321 ducats, 21 grossi, 13 pizoli and $\frac{90396}{417852}$.

To prove this, take the sum of the profits, thus:

Sebastiano gained

ducats	300	grossi	2	pizoli	8 $\frac{327456}{417852}$

Jacomo gained

ducats	321	grossi	21	pizoli	23 $\frac{90396}{417852}$

Sum of the gains

ducats	622	grossi	0	pizoli	0 $\frac{0}{417852}$

So this partnership is well and justly settled.

The third problem arranges itself as follows:

Three men, Tomasso, Domenego, and Nicolo, entered into partnership. Tomasso put in 760 ducats on the first day of January, 1472, and on the first day of April took out 200 ducats. Domenego put in 616 ducats on the first day of February, 1472, and on the first day of June took out 96 ducats.

F. 50, r.

Nicolo put in 892 ducats on the 1st day of February, 1472, and on the 1st day of March took out 252 ducats. And on the 1st day of January, [1475,], they found that they had gained 3168 ducats, 13 grossi and $\frac{1}{2}$. Required is the share of each.

In this problem, since each has taken out from his share, it is necessary to multiply each whole share by as many months as it remains untouched, as here.

```
         3)   760
         4)   616
   *[i]  3)   892
             2280
             2464
             2676
              760
              200
remainder     560(33
              616
               96
remainder     520(31
              892
              252
remainder     640(32
              560
               33
             1680
            1680
            18480
             2280
            20760
            18584
            23156
divisor     62500
```

These are the shares which are to be added to the others which arise, after subtracting the money, through multiplication of the remainder by the time, that is each by its own.

Each of these remainders should be multiplied by the rest of the time which the partnership lasts.

```
  520          640
   31           32
  520         1280
 1560         1920
16120        20480
 2464         2676
18584        23156
```

Then the shares of each are adjusted and added.

[*Nicolo is incorrectly credited with 3 months full investment when he actually kept his 892 ducats in the partnership for only 1 month. The computation proceeds under this assumption. Thus the solution does not satisfy the given conditions of the problem.]

Now arrange the rule as to Tomasso. F. 50, v.

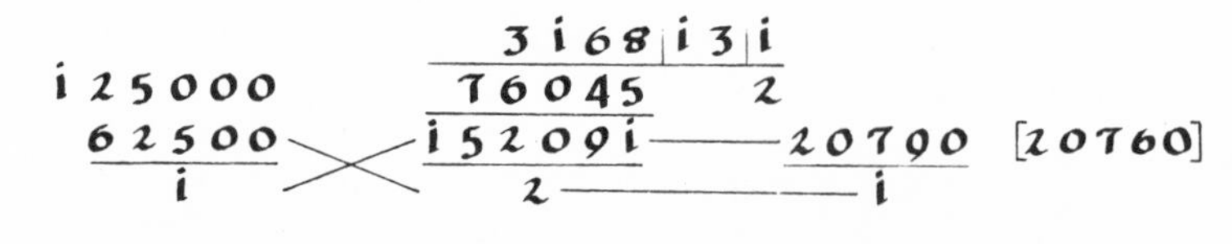

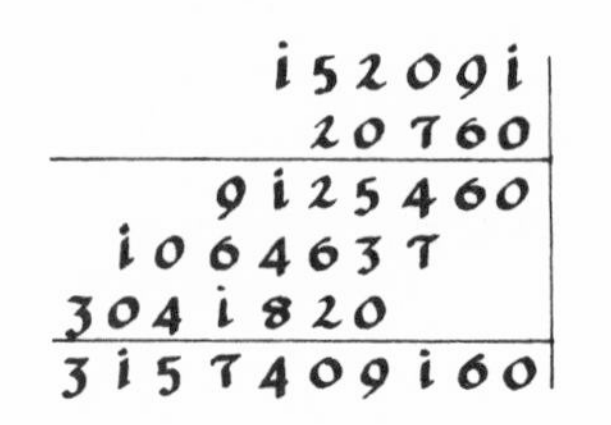

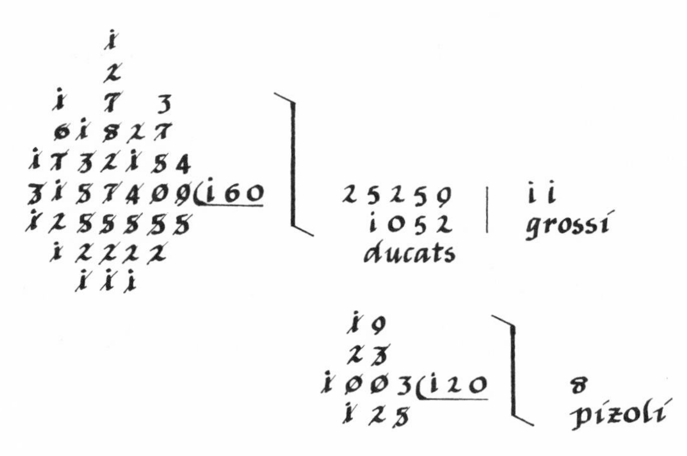

So the work is completed, and the answer is that Tomasso receives as profit 1052 ducats, 11 grossi, $8\frac{93120}{125000}$ pizoli.

Now arrange the rule as to Domenego, thus:

F. 51, r.

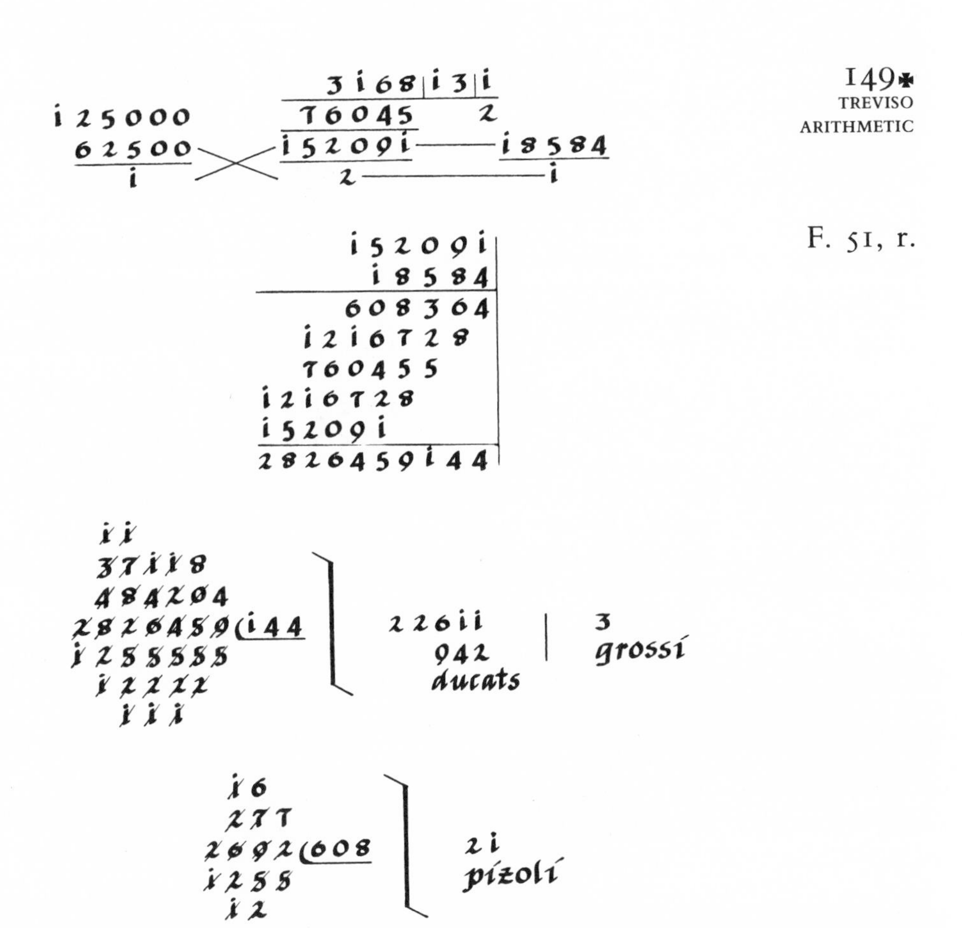

So the problem is solved, and the answer is that Domenego's share of profit is 942 ducats, 3 grossi, 21 pizoli and $\frac{67608}{125000}$.

Now arrange the rule for Nicolo, thus:

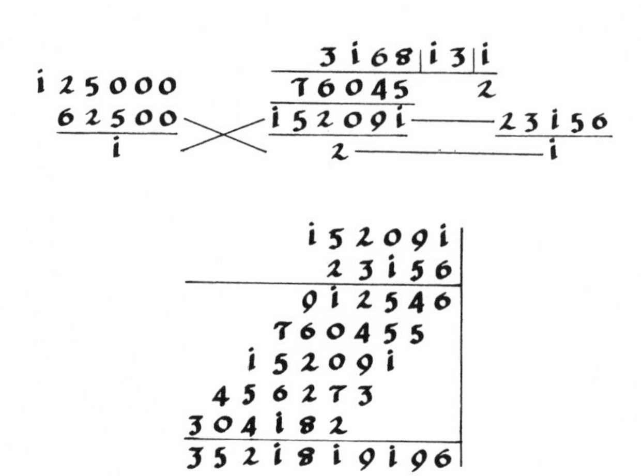

F. 51, v.

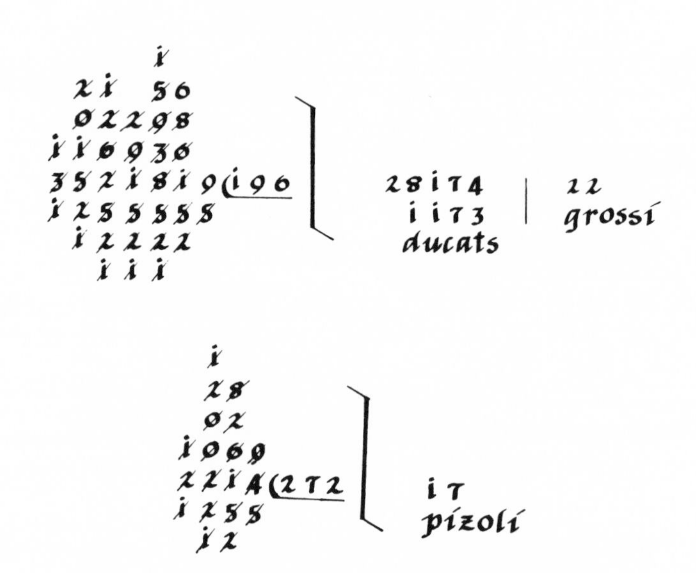

Thus the problem is solved, and the answer is that Nicolo's share of profit is 1173 ducats, 22 grossi, $17\frac{89272}{125000}$ pizoli.

To prove that the result is correct, add the profits.

Tomasso gains

ducats 1052 grossi 11 pizoli 8 $\frac{93120}{125000}$

Domenego gains

ducats 942 grossi 3 pizoli 21 $\frac{67600^{*}}{125000}$

Nicolo gains

ducats 1173 grossi 22 pizoli 17 $\frac{89272}{125000}$

Sum of the gains

ducats 3168 grossi 13 pizoli 16 1

250000 | 2
125000

*[67608]

This is the partnership settled justly without anyone being cheated.

As to the fifth [fourth] kind of problem, i.e., concerning barter.

First problem.

Two merchants wish to barter. The one has cloth at 5 lire a yard, and the other has wool at 18 lire a hundredweight. How much cloth should the first have for 464 hundredweights of wool? The rule: first find the value of the wool, thus:

F. 52, r.

```
 464
  18
----
3712
464
----
8352
```

Then arrange the rule thus:

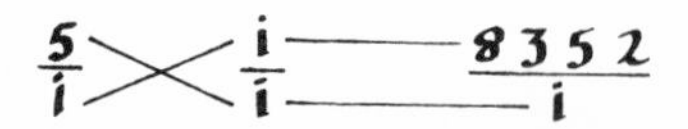

Do then as the rule requires, and you will find the result to be 1670 yards and $\frac{2}{5}$. And this way you may solve similar problems.

Second problem.

There are two merchants of whom the one has cloth worth 22 soldi a yard, but who holds it in barter at 27 soldi. The other has wool which is worth in the country 19 lire per hundredweight. Required is to know how much he must ask per hundredweight in barter so that he may not be cheated. Now arrange the rule as follows:

$$\frac{22}{1} \times \frac{27}{1} \text{———} \frac{380}{1}$$

Do then as the rule requires, and you will find that he should hold the wool at 23 lire, 6 [7] soldi and $\frac{4}{11}$ per hundredweight.

Third problem.

There are two merchants who wish to barter. One has 1 pexo of balsam worth 150 ducats. He wishes to trade this for three kinds of merchandise, namely, for wax at 5 ducats per hundredweight, sugar at 6 ducats a hundredweight, and ginger at 8 ducats a

F. 52, v.

hundredweight. He wishes the same amount of one substance as the rest of the three kinds of merchandise.
Required is how much he will have of each.
Now keep in mind the following figure:

Hundredweight of wax — ducats — 5
Hundredweight of sugar — ducats — 6
Hundredweight of ginger — ducats — 8
Divisor 19

Now arrange the rule as follows:

$$\frac{19}{1} \times \frac{100}{1} \text{———} \frac{150}{1}$$

100 times 150 makes 15000.

Dividing by 19, the quotient is 789 and $\frac{9}{19}$.
Therefore, the answer is that the man having the balsam should receive of each of the three kinds of merchandise per pexo of balsam, 789 pounds and $\frac{9}{19}$. By the same method solve all similar problems.

The fifth kind of problem above concerns the alloys of coins

First problem.

A merchant has 46 marks, 7 ounces of silver, alloyed at 7 ounces and $\frac{1}{4}$ per mark. He wishes to coin this so that it shall contain 3 ounces and $\frac{1}{2}$ of fine silver per mark. Required is to know the amount in the mixture and how much brass he must add.

In this problem you should first notice how much silver is found in the given amount of 7 ounces and $\frac{1}{4}$ per mark, and proceed by the rule of three, as follows:

F. 53, r.

If 1 mark yields 7 ounces and $\frac{1}{4}$, how much will 46 marks, 7 ounces yield? Arrange the rule thus:

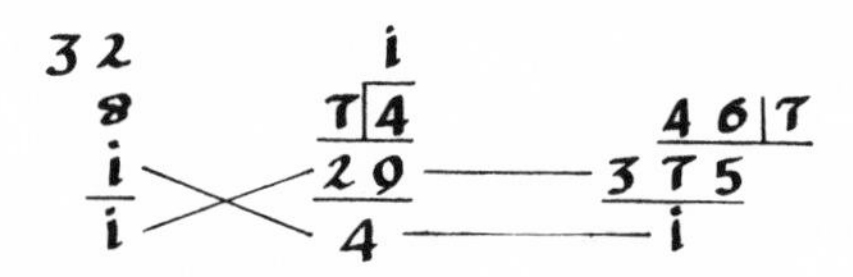

Multiplying and dividing, we find that in this amount there is 42 marks, 3 ounces, 3 qr., 13k. and $\frac{1}{2}$ of fine silver.

This being done, proceed as follows: If 3 ounces and $\frac{1}{2}$ of fine silver makes 1 mark of the abovementioned money, how much would be produced by 42 marks, 3 ounces, 3 qr., 13 k. and $\frac{1}{2}$ of fine silver? Therefore arrange this rule as follows:

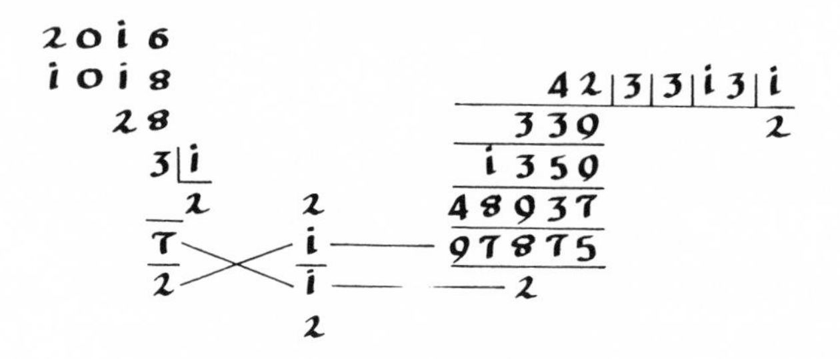

Multiplying and dividing according to the directions of the rule of 3, it appears that the money will amount to 97 marks, 0 ounces, 3 qr., 5 k. and $\frac{1}{7}$.

If you wish to know how much brass is added in the above sum, take the sum of the fineness, i.e., the 42 marks, 3 ounces, 3 qr., 13 k. and $\frac{1}{2}$, from the total weight of the coin, thus:

F. 53, v.

97 marks,	0 oz.,	3 qr.,	5 $\frac{1}{7}$ k.
42 marks,	3 oz.,	3 qr.,	13 $\frac{1}{2}$ k.
54 marks,	4 oz.,	3 qr.,	27 $\frac{8[9}{14}$ k.

So much is the brass in excess of the aforesaid 42 marks, 3 ounces, 3 qr., 13 k. and $\frac{1}{2}$. Thus the problem is solved. Note carefully to do the same in similar cases.

Second problem.

A merchant has 40 marks of silver containing 6 ounces and $\frac{1}{2}$ of fineness per mark. He has 56 marks of another kind containing 5 ounces of fineness per mark. He wishes to make these into coin containing 4 ounces and $\frac{1}{2}$ of fine silver per mark. Required is to know the amount in the mixture, and the amount of brass to be added.

In this problem you should first consider how much silver is found in the two given quantities, and first in the 40 marks, thus: If 1 mark yields 6 ounces and $\frac{1}{2}$ of silver, what will 40 marks yield? Arrange the rule thus:

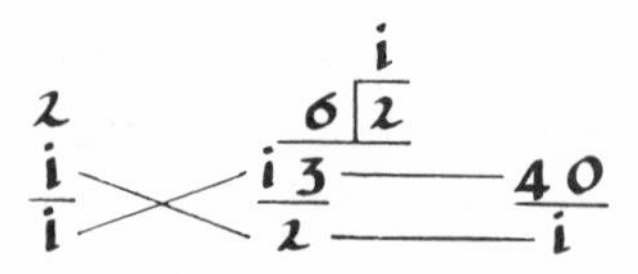

Multiply and divide, and you have 280 ounces. Then take up the 56 marks, which contains

F. 54, r.

5 ounces of silver per mark, and arrange the rule thus:

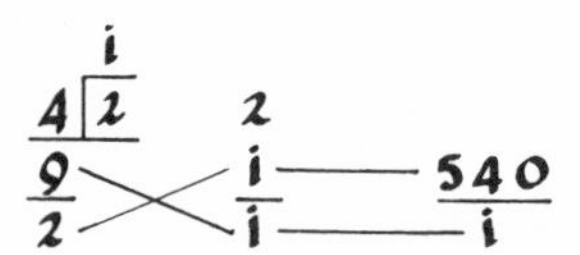

Multiply and divide and you have 280 [260] ounces to add to the 260 ounces, making 540 ounces. This is the amount of fine silver in the entire quantity. It is now necessary to reduce this to the alloy of 4 ounces and $\frac{1}{2}$ per mark. Therefore, proceed as follows: Since 4 ounces and $\frac{1}{2}$ of fine silver make 1 mark of the required money, how much will 540 ounces make? Arrange the rule as follows:

$$\begin{array}{c} 4\frac{1}{2} \quad \frac{9}{2} \times \frac{1}{1} — 540 \end{array}$$

Multiply and divide according to the directions of the rule of 3, and you will obtain 120 marks. This is the sum of all the mixture, from which sum you take 96 marks, i.e., 40 and 56 which you had in the first place; there will then remain 24. And this is the amount of brass that must be added to the aforesaid 96 marks of the two kinds to make an alloy of 4 ounces and $\frac{1}{2}$ per mark.

Third problem.

A merchant has 10 marks and 6 ounces and $\frac{1}{2}$ of silver of a fineness of 5 ounces and $\frac{1}{2}$ per mark. He has 12 marks of another kind which contains 6 ounces and $\frac{1}{2}$ per mark. He has 15 marks of another kind which has a fineness of 7 ounces and $\frac{1}{4}$ per mark.

F. 54, v.

And from all this silver he wishes to coin money which shall contain 4 ounces and $\frac{3}{4}$ of fineness per mark. Required is to know the amount in the mixture, and how much brass must be added.

This problem is solved like the preceding, and the result is 52 marks, 4 ounces, 5 k. and $\frac{416}{608}$. This is the amount of the entire mixture. From it should be subtracted 38 marks, 2 ounces and $\frac{1}{2}$, which is the sum of the three kinds of silver, and there remains 14 marks, 2 ounces, 1 qr., 5 k. and $\frac{416}{608}$. [The author's answer does not satisfy the conditions. Solving the problem as stated gives 51 marks, 6 ounces, 2 qr., 26 k. and $\frac{302}{608}$. So 14 marks, 26 k. and

$\frac{302}{608}$. So 14 marks, 26 k. and $\frac{302}{608}$ of brass must be added.] So much brass must therefore be added. In this way we may solve other similar problems.

The student has now completed the five kinds of problem finally promised by me. It remains for me to add certain accomplishments, and at the same time certain features which will be useful to him.

The rule of the two things which come together is this: That you should multiply the two terms by one another and divide the product of this multiplication by the sum of the two numbers given.

Example.

The Holy Father sent a courier from Rome to Venice, commanding him that he should reach Venice in 7 days. And the most illustrious Signoria of Venice also sent another courier to Rome, who should reach Rome in 9 days. And from Rome to Venice is 250 miles. It happened that by order of these lords the couriers started their journeys at the same time. It is required to find in how many days they will meet,

F. 55, r.

and how many miles each will have traveled. Follow the rule, thus:

Therefore, in 3 days and $\frac{15}{16}$ they will meet.

If you wish to know how many miles each has made, proceed by the rule of 3, thus: And first as to the one from Rome.

```
  112
   7         250 ——— 63
  ——   ><   ———      ——
   1          1  ——— 16
             16
```

```
   250
    63
  ————
   750
  500      [1500]
 —————
 15750
```

```
    1
   457
  15750  |  140
  11222
   111
    1
```

The one who came from Rome has made 140 miles and $\frac{5}{8}$.

Now arrange the rule for the courier from Venice.

```
  144
   9         250 ——— 63
  ——   ><   ———      ——
   1          1  ——— 16
             16
```

```
    45
   1394
  15750  |  109
  14444
   144
    1
```

Therefore, we see that the one who came from Venice to Rome has made 109 miles and $\frac{3}{8}$.

F. 55, v.

To prove the work, notice that the two together have made 250 miles thus:

	140 $\frac{5}{8}$
	109 $\frac{3}{8}$
Sum of the miles	250

The rule of the two terms which fall and come together is this: That you should multiply the two numbers by the number of paces passed over, and divide by the difference in magnitude of the two numbers.

Example.

A hare is 150 paces ahead of a hound, which pursues him. The hare covers 6 paces, while the hound covers 10. Required is to know how many paces the hound has made when he overtakes the hare.

The difference between 6 and 10 is 4, which is the divisor. Now follow the rule in this way:

	150	
	10	The hound has covered 375
	1500	paces in overtaking the hare.
Paces	375	

If you wish to prove it, find the number of paces the hare has covered, thus:

	150	
	6	
	900	The hare has covered 225 paces
Paces	225	when the hound has overtaken him.

F. 56, r.

Add to these 225 paces the start of 150 which the hare had, and the result is 375. Thus, between the distance traveled by the hare and the start he had, we have as many paces as the hound traveled. Thus the problem is finished.

Note also that similar problems are solved in this way.

A man has found a purse containing some ducats, I do not say how many. Of these he spends $\frac{1}{4}$, $\frac{1}{5}$, $\frac{1}{6}$, and 9 ducats remain. Required is the number of ducats in the purse when he found it.

These three numbers [denominators] are contained in 120, as you will find by multiplying the denominators one by the other, first the first two, then last by the result of the first two, thus: 4 times 5 are 20, and 6 times 20 are 120. Canceling 2, we have 60, in which these fractions [denominators] are also contained. Now do as follows:

The $\frac{1}{4}$			15
The $\frac{1}{5}$	of 60 is		12
The $\frac{1}{6}$			10
The sum			37

Subtract 37 from 60
37
remainder 23

Then arrange the rule thus:

$$\frac{23}{1} \times \frac{60}{1} — \frac{9}{1}$$

60
9
540

1
2
181
540 | 23
233
2

F. 56, v.

Thus it is finished, and the answer is that he found in it 23 ducats and $\frac{11}{23}$. And note carefully to proceed thus in similar cases.

I have found a purse with ducats, I do not tell you how many. I have spent $\frac{1}{3}$ and $\frac{1}{4}$ of them, and I now have 120 ducats in the purse. Required is the number of ducats in the purse when I found it.

These numbers are contained in 12, as you will see by multiplying one denominator by the other, thus: 3 times 4 are 12. Whence $\frac{1}{3}$ of 12 is 4, and $\frac{1}{4}$ is 3. Since 3 and 4 make 7, this is the number spent. Therefore, there are 5 left out of 12. Therefore, arrange the rule thus: If 5 are left from 12, from how many would 120 be left?

$$\frac{5}{1} \times \frac{12}{1} \text{———} \frac{120}{1}$$

```
 120
  12
 ---
 240
120
----
1440
```

The proof:

288 — divide by 3
   — divide by 4

```
 96
 72
---
168  amount spent
```

ducats	288
remainder	120
amount spent	168

The answer is that you found 288 ducats in the purse.

F. 57, r.

A carpenter has undertaken to build a house in 20 days. He takes on another man and says: "If we build the house together, we can accomplish the work in 8 days." Required is to know how long it would take this other man to build it alone.

In this problem we have to consider that the second man does as much work in 8 days as the first does in 12 days. Considering this, arrange the rule as follows: If 12 produces 8, what will produce 20?

12/1 — 8/1 — 20/1

13 days $\frac{1}{3}$

The answer is that the second man can build the house in 13 days and $\frac{1}{3}$.

If 17 men build 2 houses in 9 days, how many days will it take 20 men to build 5 houses?

This problem requires applying the rule of 3 twice, as may be seen here.

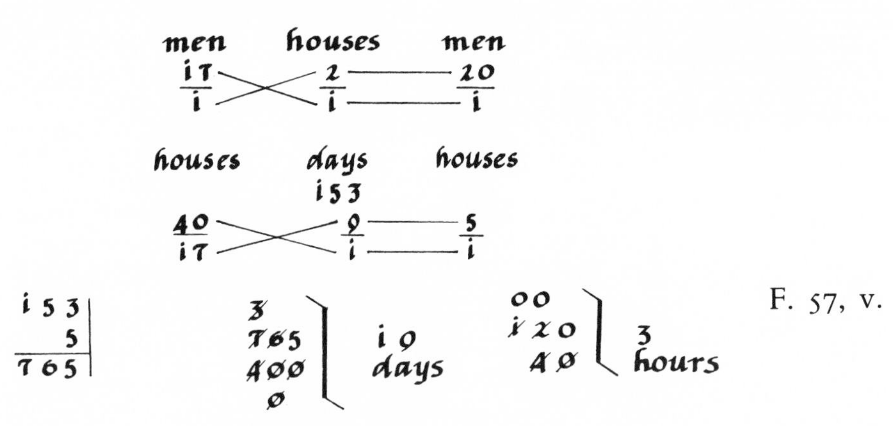

The answer is that 20 men can build 5 houses in 19 days and 3 hours.

If 3 men eat 3 loaves in 4 days, required to know in how many days 10 men will eat 12 loaves? This problem is solved (if you consider it well) like the preceding one.

The rule for finding the golden number is this: Divide by 19 the years of the Nativity of our Lord Jesus Christ which end in the year in which you seek the golden number. And paying no attention to the quotient, take the remainder and add 1 to it, and this is the golden number of the year sought.

Example.

Suppose we wish to know the golden number of the present year, viz., 1478. Do as follows:

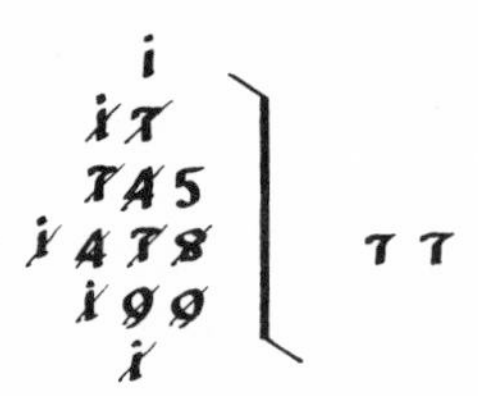

The remainder is 15. Add 1 and we have 16, and this is the golden number of 1478.

Notice that by this golden number is found the new moon in the calendar in this way: Look in a correct calendar in the month in which you wish to know when the new moon is to occur,

F. 58, r.

and when you find the golden number, the fourth day, counting from it inclusively, will be the new moon. But keep well in mind to have a calendar in which the golden number is given in its proper place.

Taking a special case, suppose you wish to know when the [new] moon will be in the present month, that is in December 1478. Now look for the aforesaid golden number, i.e., 16, and it is found at the 27th day. Begin and count from this day backwards, saying: "one, two, three, four." Therefore, since this four falls on the 24th day of the aforesaid month, namely in the vigil of the Nativity of our Lord and Saviour Jesus Christ, this shows that this day will be the new moon. And to the end that you may know further, i.e., in what day, in what hour, and in what point the moon may appear, note carefully the following rule.

Each moon has 29 days, 12 hours, and 793 points. Each hour has 1080 points. Hence, if you wish to know when the moon will be in the present month, i.e., in December of 1478, it is necessary that you should know when it was new in the month immediately preceding, that is in November. This was the 25th day, 8 hours, 408 points. This being known, place beneath these numbers, that is 25 | 8 | 408, the total duration of a moon, which lasts (as stated above) 29 days, 12 hours, and 793 points. Place these numbers one above the other, so that the days stand below days, and hours below hours, and points below points, like this:

25	8	408
29	12	793

F. 58, v.

Now add these numbers, beginning with the points, the sum of which is 1201. Now, when the sum of points exceeds the number of points in an hour (as here), write under those the number of points in an hour, which are 1080, and subtract it from the sum found. The remainder, which is 121, are the points of the present month, with the condition that you carry for the points of one hour subtracted one hour to the hours to be added, i.e., to 12, making 13, and 8 making 21, the number of hours. Also, when the number of hours exceeds the number of hours of a day, which are 24, subtract 24 from that number. The remainder represents the number of hours, with this condition, that you carry in this case a day to the number of days, 1. Then add the days, 29 with 25, making 54. Also note that when the number of days exceeds the number of all days of the preceding month (as here), where November has 30 days, subtract the days of the preceding month, i.e., 30, from the number of days added, i.e., from 54, and the remainder, 24, is the number of the day in which the moon appears. In this way we know that the moon will be new on the 24th day, on the 21st hour, and at 121 points of December, 1478, as you can understand from the following:

25	8	408
29	12	793
54		1201
30		1080
24	21	121

Note that when the month preceding the month in which you wish to know the moon has 31 days, subtract from the

F. 59, r.

sum of the days added 31, and the remainder

will be the number of the day in which the moon will appear. And when the preceding month has 30, subtract 30; and when it has 28 (as in February), subtract 28; and when 29 (as in the month of February in the bisextile year), subtract 29.

If you wish to know or find when the moon will appear in the month of January, 1479, take the number of days, hours, and points of the moon which make, or rather are made, in December, and under this, or rather, under these numbers, write (as said above) the number of days, hours, and points of the moon, namely, 29 | 12 | 793. Then add, and subtract if necessary, as above said, and so you will find as here:

24	21	121
29	12	793
54	33	914
31	24	
23	9	914

Thus the moon will be new in January, 1479, on the 23rd day, at the 9th hour, and at 914 points. In this same way you will always be able to find in what day and at what hour and point the moon will appear.

Note very carefully that you must always subtract the days of the month in which you wish to find the moon, and not of the month which you seek. So, if you wish to know the moon of the present December, subtract from the sum of the days the number of the days of the month of November. And if you wish to find the moon for next January, subtract the number of days of the month of December, and so for others.

And these two rules suffice for the new moon.

F. 59, v.

Note here concerning the conversion of pexi composed of pounds, ounces, etc., and keep in mind.

Note that to make pounds of ounces, divide the ounces by	12
And to make ounces of pounds, multiply the pounds by	12
Know that to reduce ounces to pexi, divide by	300
And to reduce pexi to ounces, multiply by	300
Note that to reduce ounces to hundredweight, divide by	1200
And to reduce hundredweight to ounces, multiply by	1200
Know that to reduce ounces to thousandweight, divide by	12000
And to reduce thousandweight to ounces, multiply by	12000
Note that to reduce pounds to pexi, divide by	25
And to reduce pexi to pounds, multiply by	25
Know that to reduce pounds to hundredweight, divide by	100
And to reduce hundredweight to pounds, multiply by	100

Note that to reduce pounds to thousandweight, divide by 1000
And to reduce thousandweight to pounds, multiply by 1000

Know that to reduce pexi to hundredweight, divide by 4
And to reduce hundredweight to pexi, multiply by 4

Note that to reduce pexi to thousandweight, divide by 40
And to reduce thousandweight to pexi, multiply by 40

Know that to reduce hundredweight to thousandweight, divide by 10
And to reduce thousandweight to hundredweight, multiply by 10

F. 60, r.

Note that to reduce pounds of 12 pounds of 18 ounces, multiply by 12 and divide by 18.
Know that to reduce pounds of 18 ounces to pounds of 12 ounces, multiply by 18 and divide by 12.

Note here concerning the question of prices known, and by this to understand other prices not known.
Note that knowing the value of the ounce, and wishing to know the value of the pound, multiply the value of the ounce by 12, and the result will be the value of the pound.
And note that knowing the value of the pound, and wishing to know the value of the ounce, divide the value of the pound by 12, and the result will be the value of the ounce.
Note that knowing the value of the ounce, and wishing to know the value of the pexo,

which contains 25 pounds, multiply the value of the ounce by 300 and the result will be the value of the pexo.

And note that knowing the value of the pexo, and wishing to know the value of the ounce, divide the value of the pexo by 300, and the result will be the value of the ounce.

Note that knowing the value of the ounce, and wishing to know the value of the hundredweight, multiply the value of the ounce by 1200, and the result will be the value of the hundredweight.

And note that knowing the value of the hundredweight, and wishing to know the value of the ounce, divide the hundredweight by 1200, and the result will be the value of the ounce.

Note that knowing the value of the ounce, and wishing to know the value of the thousandweight, multiply the value of the ounce by 12000, and the result will be the value of the thousandweight.

And note that knowing the value of the thousandweight, and wishing to know the value of the ounce, divide the value of the thousandweight by 12000, and the result will be the value of the ounce.

Note that knowing the value of the pound, and wishing

F. 60, v.

to know the value of the pexo, multiply the value of the pound by 25, and the result will be the value of the pexo.

And note that knowing the value of the pexo, and wishing to know the value of the pound, divide the value of the pexo by 25, and the quotient will be the value of the pound.

Note that knowing the value of the pound, and wishing to know the value of the hundredweight, multiply the value of the pound by 100, and the result will be the value of the hundredweight.

And note that knowing the value of the hundredweight, and wishing to know the value of the pound, divide the value of the hundredweight by 100, and the quotient will be the value of the pound.

Note that knowing the value of the pound, and wishing to know the value of the thousandweight, multiply the value of the pound by 1000, and the result will be the value of the thousandweight.

And note that knowing the value of the thousandweight, and wishing to know the value of the pound, divide the value of the thousandweight by 1000; the quotient will be the value of the pound.

Note that knowing the value of the pexo, and wishing to know the value of the hundredweight, multiply the value of the pexo by 4, and the result will be the value of the hundredweight.

And note that knowing the value of the hundredweight, and wishing to know the value of the pexo, divide the value of the hundredweight by 4, and the quotient will be the value of the pexo.

Note that knowing the value of the pexo, and wishing to know the value of the thousandweight, multiply the value of the pexo by 40, and the result will be the value of the thousandweight.

And note that knowing the value of the thousandweight and wishing to know the value of the pexo, divide the value of the thousandweight by 40; the quotient will be the price of the pexo.

Note that knowing the value of the hundredweight, and wishing to know the price of the thousandweight, multiply the price of the hundredweight by 10, and the result will be the value of the thousandweight.

F. 61, r.

Know that knowing the value of the thousandweight, and wishing to know the value of the hundredweight, divide the value of the thousandweight by 10, and the quotient will be the value of the hundredweight.

Note here below concerning the values of fractional parts, viz., of ounces, of pounds, of pexi, and of hundredweight, and of thousandweight. Therefore keep in mind as to the multiplication or divisions that may come.

Note that for every penny that the ounce is worth, the pound is worth 1 soldo; the pexo 1 lira, 5 soldi; the hundredweight 5 lire; the thousandweight 50 lire.

Know that for every soldo that the ounce is worth, the pound is worth 12 soldi; the pexo 15 lire; the hundredweight 60 lire, the thousandweight is worth 600 lire.

Note that for every penny that the pound is worth, the ounce is worth $\frac{1}{12}$ of a penny; the pexo 2 soldi, 1 penny; the hundredweight is worth 8 soldi, 4 pence; the thousandweight is worth 4 lire, 3 soldi, 4 pence.

Know that for every soldo that the pound is worth, the ounce is worth 1 penny, the pexo 1 lira, 5 soldi, the hundredweight 5 lire, and the thousandweight 50 lire.

And know that for every lira that the pound is worth, the ounce is worth 1 soldo, 8 pence, the pexo 25 lire, the hundredweight 100 lire, and the thousandweight 1000 lire.

Note that for every penny that the pexo is worth, the ounce is worth $\frac{1}{300}$ of a penny, and the pound is worth $\frac{1}{25}$ of a penny, the hundredweight 4 pence, and the thousandweight 3 soldi, 4 pence.

Know that for every soldo that the pexo is worth, the ounce is worth $\frac{1}{25}$ of a penny, the pound is worth $\frac{12}{25}$ of a penny, the hundredweight is worth 4 soldi, and the thousandweight is worth 2 lire.

And know that for every lira that the pexo is worth, the ounce is worth $\frac{4}{5}$ of a penny, the pound 9 pence and $\frac{3}{5}$, the hundredweight 4 lire, the thousandweight 40 lire.

Note that for every penny that the hundredweight is worth, the ounce is worth $\frac{1}{1200}$ of a penny, and the pound is worth $\frac{1}{100}$ of a penny, the pexo is worth $\frac{1}{4}$ of a penny, the thousandweight is worth 10 pence.

F. 61, v.

Know that for every soldo that the hundredweight is worth, the ounce is worth $\frac{1}{100}$ of a penny, and the pound is worth $\frac{3}{25}$ of a penny, the pexo is worth 3 pence, the thousandweight is worth 10 soldi.

And know that for every lira that the hundredweight is worth, the ounce is worth $\frac{1}{5}$ of a penny, and the pound is worth 2 pence and $\frac{2}{5}$, the pexo is worth 5 soldi, and the hundredweight is worth 10 lire.

Note that for every penny the thousandweight is worth, the ounce is worth $\frac{1}{12000}$ of a penny, and the pound is worth $\frac{1}{1000}$ of a penny, the pexo is worth $\frac{1}{40}$ of a penny, the hundredweight is worth $\frac{1}{10}$ of a penny.

Know that for every soldo that the thousandweight is worth, the ounce is worth $\frac{1}{100}\left[\frac{1}{1000}\right]$ of a penny, and the pound is worth $\frac{3}{250}$ of a penny, the pexo is worth $\frac{3}{10}$ of a penny, the hundredweight is worth 1 penny and $\frac{1}{5}$ of a penny.

And know that for every lira that the thousandweight is worth, the ounce is worth $\frac{1}{50}$ of a penny, and the pound is worth $\frac{6}{25}$ of a penny, the pexo is worth 6 pence, the hundredweight is worth 2 soldi.

And now, my dearest friends, the work is finished. With great affection my request is granted. By as much study as you have given the work, by so much has it appealed to your ardent desires. I do not doubt that it will bring back to you much fruit. Not that by such sacrifice, however, can we lead everyone to be learned or expert in this practice (since to them such teachings may not be necessary), but only to such as you who are desirous of education. And therefore according to your wish, if not wholly at least in part, and corresponding to my effort by you so graciously received, I promise you the same gratifying usefulness.

Finis.

F. 62, r.

This is the register of the quartos of the present work. [Then follows the initial words of the successive folios of each quarto, the first beginning "Incommincia," the second "tia. de la," etc. They are grouped in fours, this being a quarto book. There are no words for folios 5–8, these being parts of the first four sheets. Folio 9 begins with "che. 2. o di," and so on.]

What availeth virtue to him who does not labor? Nothing.

At Treviso, on the 10th day of December, 1478.

✠ 3
An Introduction to the New Knowledge

"Incommincia. . ."[1]

The author's opening comments indicate a strong analogy with a reckoning school situation; although no formal mention is made of a school per se, the indication is that the author is a *maestro d'abbaco* (master of computing—a reckoning master), preparing instruction for "youths" who look forward to mercantile pursuits.[2] Through his writing he intends to teach them the commercial art of reckoning *(larte de labbacho)*. By the time of the late fifteenth century, the abacus had been associated with arithmetic for such a long period that the word "abacus" served as a synonym for computing; actually, in the Italian of this era, *abbaco* could refer to numerals, the practice of arithmetic, or an arithmetic book, depending on the context of its usage.[3]

While the text is mainly applied in nature, some theoretical and philosophical considerations are incorporated into the instruction. The statement "All things which have existed since the beginning of time have owed their origin to number," is a quotation from Aristotle via Boethius[4] and found its way into the writings of many early mathematicians, e.g., Sacrobosco.[5] The concept of a natural number is defined further when the reader is told that a number is a collection of units, the smallest [natural] number being two.[6] One, itself, was not considered a number, but rather the builder of numbers. This principle was adhered to by all writers of early Renaissance arithmetics and advocated in various forms:

> Number is a multitude composed of units. Aristotle says, if anything is infinite, number is, and Euclid in the third postulate of the seventh book says that its series can proceed to infinity, and that it can be made greater than any given number by adding one.[7]

> Unity is the beginning of all number and measure, for as we measure things by number, we measure number by unity.[8]

> Unity is not a number, but the source of number.[9]

> Unity is the basis of all number, constituting the first in itself.[10]

The Greeks and many other ancient peoples followed this precept.[11] A student of modern mathematics can appreciate the similarity of this theory with that of Guiseppe Peano (1858–1932) in his development of the natural numbers using the concept of successor elements in a given set.[12] However, the fact that one, itself, is excluded from the realm of number at this period of history may not be due directly to philosophical or mathematical considerations inherited from antiquity, but perhaps rather to an oversight in classical translation. Necomachus (c.100 A.D.), in his consideration of Greek number theory, had written that unity was not a *polygonal* number; Boethius inadvertently mistranslated this phrase leaving out the modifier "polygonal" and merely stated that one was not a number.[13] Later writers simply transcribed Boethius on this point without questioning the issue.

Numbers were considered discrete objects—no conception of a continuum or associating numbers with points on a line had yet arisen. While the Greeks had associated numerical magnitude with the length of a line segment, this subtlety had, at this time, not been resurrected by Renaissance mathematicians.[14]

Following a medieval practice, numbers were divided into three categories: the nine *digiti* or "fingers" from 1 to 9; the *articuli* or "joints" designating multiples of ten, such as 10, 50, 650, etc., and the *numeri compositi* or "composite numbers" which were those formed by joining the previous two classes, i.e., 24, 753, etc. The terms *digiti* and *articuli,* in such a mathematical context, owe their origins to early finger counting. They later became incorporated as technical terms in the operation of an abacus and finally, by the thirteenth century, ended up as descriptive number classifiers. *Numeri compositi* refers to numbers composed of *digiti* and *articuli*. The

Treviso author replaced the term *digiti* by "simple" but retained the remaining categories of the classification system.[15]

Five operations, actually referred to as acts, *atti,* are to be dealt with in the practica: numeration, addition, subtraction, multiplication, and division. The exact number of mathematical operations, vicariously called parts or species in early arithmetic books, varied from author to author; for example, Sacrobosco in his writings gives nine: numeration, addition, subtraction, mediation, duplication, multiplication, division, progression, and square and cube root extraction, whereas de Villdeau presents seven. Neither was the order of presenting the operations universally agreed upon by authors of this time—the *Treviso's* order would win contempory pedagogical approval, but the order employed by such authors as Rabbi ben Ezra (c. 1140) and Fibonacci (multiplication, division, addition, subtraction, . . .) would certainly be subject to question.

These early concepts of a mathematical operation are quite broad and certainly differ from the present-day notion of this term. Apparently for the writers of medieval and Renaissance arithmetic books, a mathematical operation could be any specific mathematical technique or procedure they considered of special importance and thus worthy of study. For example, numeration required both the learning of a new system of symbols and a more thorough understanding of place value than had previously been necessary in mathematical work. Thus, in itself, numeration comprised a theory and was classified as a mathematical operation. Today, the term "operation" in mathematics denotes a special type of function. In considering contemporary arithmetic, one refers to "the four basic operations," namely addition, subtraction, multiplication, and division, all of which satisfy the modern functional connotation for an operation.

Numeration

The first operation to be considered is numeration, which is defined in a rather modern vein as the representation of numbers by symbols.[16] To fully appreciate the task the *Treviso's* author is undertaking in this section, one should understand the level of acceptance for the "Hindu-Arabic"[17] numeral system that existed in Europe at this time. The new numerals had been known in Europe from about 1000 A.D. yet they had not been universally accepted for use. Computing and the techniques of arithmetic still centered around the manipulation of counters and recording one's results with Roman numerals. There was a certain social status and prestige associated with the use of a counting table. Shakespeare in his works makes frequent reference to the use of counters; as the clown from *The Winter's Tale* notes:

> Let me see. Every 'leven wether tods;
> every tod yields pound and odd shilling;
> fifteen hundred shorn, what comes the wool to?
> . . . I cannot do't without counters.[18]

In a very true sense, the use of Roman numerals and counting boards had become an institution, one vested literally in the hands of a select few. Since the use of algorithms associated with the Hindu-Arabic numerals, if popularized, could be easily learned and performed without elaborate equipment, its knowledge presented a definite threat to the well-being and livelihood of established computers; therefore, it was resisted.[19] Even in Italy, where a heightened sense of mathematical awareness existed, acceptance of the Hindu-Arabic numerals was slow in coming. A *Statuto dell' Arte di Cambio* issued in 1299 by the City Council of Florence required that accounting book entries be in Roman numerals.[20] In 1348, the University of Padua required that lists of its books have their prices affixed in Roman numerals—*"non per cifras, sed per literas claras"* (not by figures, but by clear letters).[21] Ostensibly, such regulations were intended to reduce instances of fraud, as it was felt the new numerals were vulnerable to tampering. A Venetian manual on bookkeeping explains:

> the old figures alone are used because they cannot be falsified as easily as those of the new art of computation, of which one can, with ease, make one out of another, such as turning the zero into a 6 or a 9, and similarly many others can also be falsified.[22]

Legal courts gave precedence to documents using Roman numerals over those bearing the new numerical symbols. In some instances, reckoning masters were simply forbidden to use Hindu-Arabic numerals: "moreover, the master calculators are to abstain from calculating with digits," decreed a Frankfurt ordinance of 1494.[23] But, despite the sanctions against their use, the Hindu-Arabic numbers gradually became accepted and admired for their recording and computational efficiency by the merchant houses of Italy and other countries. The earliest known practical instructions for learning bookkeeping employing the new numerals was offered in the writings of the Florentine Paolo Dagomari dell' Abaco (c.1218–1374).[24] Medici account books preserved in the Selfridge Collection of Harvard's Graduate School of Business Administration show the gradual transition from Roman to Hindu-Arabic numerals that took place in Italy during the course of the fifteenth century.[25] In examining the account books of the Venetian merchant family Barbarigo, Frederic Lane noted that the first Barbarigo ledger to have all entries written in Hindu-Arabic numerals was for the years 1496–1528.[26] He did, however, find an account of one Jacomo Badoer who recorded his business dealings in Constantinople for the period between September 2, 1436 and February 26, 1439 using Arabic numerals.[27] Eventually, the new numerals became so closely associated with the transactions of merchants that they were popularly called *figura mercantesco,* mercantile figures.[28] Although in some instances, such as the minting of coins,[29] northern Europeans preceded the Italians in the use of the new numerals, in general, it can be said that by the early fifteenth century, Italy was well ahead of the rest of Europe in the art of using of Hindu-Arabic numerals in record keeping and computation. In fact, by the time of the *Treviso's* printing, the physical forms of the numerals used in Italy had evolved into the symbols known today[30] (see Table 3.1). This situation was not the same throughout Europe. Thus, the *Treviso's* author was offering new and somewhat controversial knowledge to his charges.

Two peculiarities of early printing are carried into the translation: numerals are sometimes offset from the text by the use of periods; and either because of economy or a real lack of sufficient numbers of type fonts for *1*, *i*'s have been substituted for the unit numeral.

Zero is referred to merely as a symbol to be used in conjunction with digits to express a number. By itself, it was literally said to mean nothing, i.e., *nulla, nulle, rein,* but written to the right of a digit it became a placeholder.[31] The table of F2r presents a scheme whereby the zero's function as a placeholder and the role it plays in increasing the value of a number is

Table 3.1 Evolution of Numerals

1	2	3	4	5	6	7	8	9	0	Year
[illegible]	[illegible]	[illegible]	[illegible]	[illegible]	[illegible]	[illegible]	[illegible]	[illegible]		976
[illegible]	[illegible]	[illegible]	[illegible]	[illegible]	[illegible]	[illegible]	[illegible]	[illegible]	[illegible]	1197
[illegible]	[illegible]	[illegible]	[illegible]	[illegible]	[illegible]	[illegible]	[illegible]	[illegible]	[illegible]	1275
[illegible]	[illegible]	[illegible]	[illegible]	[illegible]	[illegible]	[illegible]	[illegible]	[illegible]	[illegible]	1360
[illegible]	[illegible]	[illegible]	[illegible]	[illegible]	[illegible]	[illegible]	[illegible]	[illegible]	[illegible]	1442
i	2	3	4	5	6	7	8	9	0	1478 Treviso

effectively demonstrated. Such a graphic scheme would certainly appeal to students familiar with counting-board techniques. The idea of positional value was considered the most difficult concept to convey in early teaching of the Hindu-Arabic system.

Writers of commercial arithmetics employed various visual schemes to convey the association of positional notation with number size. Perhaps the most commonly-used scheme was that of a *castellucio,* or "little castle." This was a columnar array of digits listing numbers of varying magnitudes with the smallest, i.e., two digits, at the top and numbers in the tens or hundreds of thousands at the bottom. In the mind of the medieval writer, the array seemed to silhouette the turret of a castle and he outlined and embellished the diagram as such.

By the visual scheme given in F2v, place value is extended through a thousand million. At this period of time, arithmeticians were just beginning to cope with number values in excess of a million. The *Treviso's* table uses no number name higher than a million. The stem of the word "million" lies in the Latin *mille,* a thousand—a million being considered a large thousand. Chuquet's manuscript of 1484 supplies us with the first appearance of the terms billion, trillion, etc.[32]

Addition

The use of specific symbols to designate our four basic operations of arithmetic did not come into being until about the middle of the sixteenth century (see Table 3.2). Up until that time, a student had to determine the operation to be performed from the context of a given problem situation. Just as a modern-day mathematics teacher might advise his or her students to seek out "key words" in setting up solution procedures for word problems, the *Treviso's* author points out the important operational words: "and" for addition; "from" for subtraction; "times" for multiplication, and "in" for division.[34] Attention is called to the binary nature of the operations in question and the fact that the operations as used are closed in the set of positive numbers. The numbers employed in fifteenth-century commercial reckoning were the positive real numbers, therefore no concern was given to negative values. In this limited perspective of applying

Table 3.2 Origins of Conventional Operational Symbols in Arithmetic[33]

Operation	Symbol	First introduced
Addition	+	1489, Johann Widman, *Behede und hubsche Rechenung auff allen kauffmanschafft* (It has been theorized that +, − were symbols used to denote excess and deficiency in the commercial task of filling kegs.)
Subtraction	−	1481, Dresden manuscript; Widman (1489)
Multiplication	×	1631, William Oughtred, *Clavis mathematicae*
Division	÷	1659, Johann Heinrich Rahn, *Teutsche Algebra* (Zürich)

the four operations, the author's assertion of closure is correct.

Addition is defined as the union or joining together of several numbers. While early writers agreed on a general operational definition of addition, the standardization of terms was still some way off. Our author uses *iongere* (join) for "add" throughout his work, but John of Luna (c. 1140) referred to the operation as *aggregation,* and a French algorithm of 1275 uses *assamble.* Frequently, the word summation was used as a synonym for addition, as in Rudolff (1526) or Stifel (1545).

Commentators on the form and content of early arithmetics have puzzled over the omission of tables of basic addition facts.[35] It is true that early writers seldom included such tables in their books, but it must be remembered what kind of audience these books were intended to serve. Students of reckoning masters were adolescents, probably between twelve to sixteen years of age, who had experienced some previous basic education. Certainly, in being expected to use the books, they could read—the mark of an educated man in the fifteenth century. In achieving this basic education they were also exposed to the rudiments of arithmetic and would know elementary addition and multiplication facts. The *Treviso* author takes this for granted in his presentation; for example, in the explanation of place value in the new numeral system, he uses multiplication, although he has yet to formally define and explain it. Still, his general heuristic approach indicates an awareness of the necessity of organizing knowledge in a sequential and purposeful manner. Such a teacher would not omit such fundamental knowledge as the basic facts of addition had they not already been learned.

The development of the addition algorism in the *Treviso Arithmetic* is one of the first examples of the modern-day form. Once again, a limited realm of terminology causes some difficulties in communication. Early theoretical arithmetics spoke of *numeri addendi,* numbers to be added. As early as 1520, Gemma Frisius shortened this phrase to *addendi,* from which evolved the term "addend." Possibly incorporating a practice inherited from abacus use of counter addition, the author urges his students to order their addends with the larger addend in the uppermost position. This strategy also helps to emphasize and preserve the place value identity of the computational columns. Addition is then carried on between two numbers from right to left in the array and from the bottom to the top of columns, with the result recorded at the bottom of the array. A space rather than a line separates the resulting sum from the addends. The sum is further distinguished by the accompaning word *summa.* Where the sum of the entries in a particular column exceeds ten, the student is advised to write down the entry of lower order and carry the leftmost digit to the next column. Clearly, the physical concept of "carrying," *portare,* a number over to the next column owes its origin to the abacus, where an excess of counters on one column or line would necessitate a physical transferring or "carrying" of counters to a higher-order position. In this arithmetic, the number carried over is added to the bottommost entry of the left adjacent column, where addition again begins upwards. Not all early authors used this format; some performed addition from left to right and wrote the sum at the top or side of the addend array.[36]

Two methods are offered as checks for addition: the subtraction of one addend from the sum and the casting out of nines.[37] Subtraction is readily realized to be the inverse of addition—it can undo the results obtained in addition. While this method is suggested for a two-addend problem, no consideration is given for dealing with a problem with a larger number of addends. The method of casting out nines is very old in the history of arithmetic. Avicenna (978–1036) speaks of it as "the method of the Hindus." Transmitted from India by Arab merchants, it became popular in medieval Europe even before its "fellow traveler," the Hindu-Arabic numeral system. The acceptance of the use of such a method to assert the correctness of computation in both Eastern and Western cultures was due to the peculiarities of the computational processes used. In the East, calculations made on a dust board or sand table were erased in the course of reaching a solution; similarly, in the West, abacus or counting-table procedures eliminated primary entries in advancing towards an answer. The casting out of nines provided a check on one's work, depending only on a knowledge of the final answer and the initial problem—knowing the intermediate result was not necessary. The method was incorporated into Western mathematics for well over 800 years. All reckoning manuals of this period provided instruction on the casting out of nines; Pacioli refers to it as *"corrente mercantoria e presta"* ("appropriate for mercantile needs"), and Scheubel (1545) and Tartaglia (1556) actually gave tables of multiples and excesses of nines for their reader's reference. While at this place in the practica the author seems to imply a certainty in the validity of "proof by nines", later in the work he becomes more cautious in this assumption.[38]

The remainder of the discussion on addition is devoted to developing ability in the general use of the algorithm. Addition in more mathematically complex problems is considered: finding the sum of two two-digit numbers, three-digit numbers and four-digit numbers, a five-digit number and a four-digit number, and finally to applications in commercial arithmetic. The commercial arithmetic problems further involve the conversion of monetary units: lire, soldi, piccoli, *(pizoli),* ducats, and grossi. (A discussion of Venetian monetary conversion is given in Chapter 7.)

It is disturbing and perhaps will remain a moot question as to why the author only considers addition involving two numbers. Certainly, the commercial arithmetic of this period requires the summing of columns of figures exceeding two entries in length, and it would seem that some practice would be provided on this aspect of addition. Arithmetics prior to the fifteenth century, with the exception of Leonardo of Pisa's (1202) work, only considered problems of two addends, and the *Treviso* seems to follow this principle.[39]

Subtraction

Subtraction is also presented as a binary operation in which the larger number numeral is placed above the smaller number numeral and the difference of the two numbers is extracted. Such a procedure avoided the embarrassment of having to justify the existence of negative numbers.[40] The author does not use the accepted Latin terms *minuend* and *subtrahend,* which were abbreviations for *numerus minuendus* and *numerus subtrahendus,* but translates the complete Latin phrases, "number to be diminished" and "number to be subtracted," respectively. Some writers merely called them "larger" and "smaller" numbers; for example, Tonstall (1522) referred to the number pairs used in subtraction as *numerus superior* (larger number) and *numerus inferior* (smaller number). Similarly, a 1586 Italian translation of Clavius also refers to the subtraction entries as *numero superiore* and *numero inferiore*. The difference obtained in the *Treviso*'s problem is termed the remainder, *lo resto*. Latin writers used more extensively descriptive terms to designate the difference: Tinalus (1515), *numerus residuus,* Clavius (1583), *differentia siue excessus*. Unlike the addition examples, a bar is used in subtraction to separate the problem from the answer.[41]

In supplying directions for the actual subtraction process for instances where the particular subtrahend entry exceeds its corresponding minuend, the author uses the method of complementary subtraction where, if x is the minuend entry and y the subtrahend, then

$$x - y = x + (10 - y) - 10$$

the resulting $x + (10 - y)$ will be recorded in the column immediately under consideration and the -10 will result in a 1 being added to the adjacent left-hand element in the subtrahend. Complementary subtraction was one of the three popular methods in use at this time.[42] It was transmitted into Europe from Hindu sources, e.g., Bhaskara (c. 1150) used it in his *Lilavati*. Some other European authors of this period who advocated this method of subtraction were Widman (1489), Tonstall (1522), Pacioli (1494), Tartaglia (1556), and Riese (1522).

Rather than gradually introducing the reader to illustrative problems of varied degrees of conceptual difficulty, the author chooses, in his very first problem, to demonstrate all three subtractive situations, i.e., $x - y$ where $x > y$, $x = y$, and $x < y$:

$$\begin{array}{ccc} 4 & 5 & 2 \\ \underline{3} & \underline{4} & \underline{8} \\ 1 & 0 & 4 \end{array}$$

In the right-hand column we encounter the situation where $x < y$ and complementary subtraction is performed:

$$2 + (10 - 8) - 10 = 4 - 10$$

the 4 is recorded in this column and a 1 added to the following subtrahend entry $4 + 1 = 5$, now $x = y$ and the difference is 0, finally $x > y$, $4 - 3 = 1$ and the remainder is found to be 104.

The subtraction problems involving monetary units probably presented some difficulty to the student as the monetary units mentioned did not have a fixed base, such as ten. The major monetary units of the time were the gold ducat and the silver grosso. While the ducat was primarily a unit of commercial exchange, the grosso became the basis for daily transactions and a system of monetary units developed around it.

In applying the principle of complementary subtraction, the computer is faced with the problem of using a varying radix as he moves from one monetary unit to another. Thus, in the sixth problem, the author states that the radix, *el rezimento,* of *piccoli* is twelve, that is, 12 *piccoli* = 1 *soldo*. When working with *ducats, grossi,* and *piccoli,* as in the seventh problem, we are told that the radix of the *piccoli* is 32, i.e., 1 *grosso* = 32 *piccoli,* and that of the *grossi* is 24, i.e., 1 *ducat* = 24 *grossi*. The fact that the radix of the ducats is given as ten indicates that they are the largest monetary unit in the problem, and their base is merely the base of the number system, 10. Finally, the reader is exposed to a problem where four different radices are required—a true test of understanding.

Two methods of checking the correctness of the answer are again given; however, bias for the use of the inverse operation is expressed. In fact, on F10v, the author questions the veracity of the "casting out of nines" method of proof. In further advocating and reaffirming the use of inverse operations in proving results in both addition and subtraction, the author returns to his two original addition examples. This practice of referencing new concepts with ones already considered indicates a spiral approach to teaching and is pedagogically attractive.

Some Preliminary Impressions

From the complete reading of the *Treviso Arithmetic* and the discussion already undertaken on specific aspects of the work, it is obvious that the practica is a manual of the briefest form and presents little else than what the author considers the fundamentals of his craft. Two reasons can easily be given to substantiate this decision: printing was costly, forcing an economy of space, and the book was prepared as an adjunct to either guided instruction by a reckoning master or diligent self-study. A minimum of illustrative problems is given; rather, the author's efforts at instruction are based on rhetorical explanations and no pages of drill exercises are provided, as it was expected that those using the book would, themselves, undertake required practice. Indeed, the author makes it quite clear that he expects the student-readers to do supplementary work in order to obtain a fuller understanding of the material, "since, if you have an interest in study, you will understand clearly the method and manner from what has already been given".[43] "Do" is a characteristic feature either implicitly or explicitly given in the text. This principle of "learning by doing" was adhered to by most early writers of arithmetics. Latin arithmeticians used to write *"Fac ita,"* "do it thus," and the Germans, *"Thu ihm also,"* "do it as before."

✠ 4 Multiplication

> To understand this [multiplication] it is necessary to know that to multiply one number by itself or by another is to find from two given numbers a third which contains one of these numbers as many times as there are units in the other.[1]

This is how the *Treviso Arithmetic* defines multiplication. This definition was common in the early European works on arithmetics; for example, *The Crafte of Nombrynge* (c. 1300) states:

> multiplicacion is a bryngynge to-geder of 2 thynges in on nombor, the quych on nombor contynes so mony tymes on, howe monv tymes there ben vyntees in the nowmbre of that 2.[2]

This definition is slightly circular, in that in its explanation it refers to the process of division; however, its essence lies in the realization that multiplication, *moltiplicare,* may be viewed as repeated addition.

Multiplication is further clarified as being binary in nature. Having previously established a general format in setting up the members of a computational algorism, the author uses it once again in requiring that the larger of the two multiplication entries be used as the multiplicand and the smaller as the multiplier. This practice appears to be by convenience and convention, as he does go on to point out that the process of multiplication is commutative. The listing of basic multiplication facts that follow adhere to this format. Just as a modern-day teacher, when introducing students to the operation of multiplication, will demonstrate its basic mathematical properties (closure, commutativity, etc.) and use these properties in developing a table of basic multiplication facts in such a manner that the students' memory load is reduced, so, too, Italian Renaissance arithmetic teachers employed similar devices.

Tables of basic multiplication facts were presented in three ways in early arithmetic books: in a column, in a square array, and in a triangular array. The *Treviso* uses a column arrangement which seemed to be preferred by the commercial arithmetic authors of the time; similar configurations were employed by Pacioli (1494), Pellos (1492), Borghi (1484), Huswirt (1501), and many others. The Italians evidently obtained the concept of column arrangements for multiplication facts from Eastern sources.[3] Use of a square array was more commonly found in the noncommercial texts, particularly those based upon Boethian tradition. The square array was popularly called a Pythagorean table—*table de Pythagore, tabula Pythagorica, tavola Pitagorica,* etc.—

although any historical connection with Pythagoras or the Greeks is extremely doubtful (see Figure 4.1). The remaining display of multiplication facts, a triangular arrangement, was not as popular as the previous two, but was employed by some authors (e.g., Widman [1489] and Gemma Frisius [1540];. See Figure 4.2. It probably was transmitted to Europe from Arab sources.[4]

The author of the *Treviso Arithmetic* lacks specific terms in discussing the various elements of the multiplication algorism. He refers to them in a most general way as the number multiplied, *el nũero de fir moltiplicator* (multiplicand) and the multiplying number, *el nũero moltiplicatore* (multiplier). Latin writers

1	2	3	4	5	6	7	8	9	10
2	4	6	8	10	12	14	16	18	20
3	6	9	12	15	18	21	24	27	30
4	8	12	16	20	24	28	32	36	40
5	10	15	20	25	30	35	40	45	50
6	12	18	24	30	36	42	48	54	80
7	14	21	28	35	42	49	56	63	70
8	16	24	32	40	48	56	64	72	80
9	18	27	36	45	54	63	72	81	90
10	20	30	40	50	60	70	80	90	100

Figure 4.1 Rectangular table of multiplication facts. Tonstall (1522).

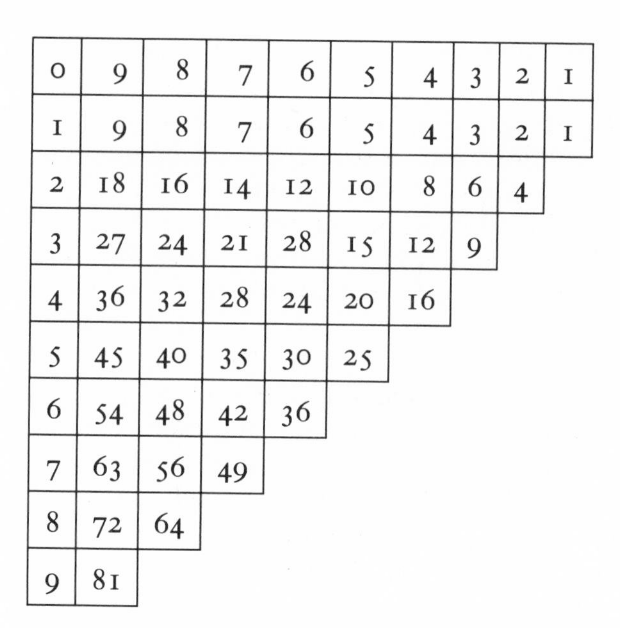

0	9	8	7	6	5	4	3	2	1
1	9	8	7	6	5	4	3	2	1
2	18	16	14	12	10	8	6	4	
3	27	24	21	28	15	12	9		
4	36	32	28	24	20	16			
5	45	40	35	30	25				
6	54	48	42	36					
7	63	56	49						
8	72	64							
9	81								

Figure 4.2 Triangular table of multiplication facts. Cirvelo (1513).

called these components *numerus multiplicandus* and *numerus multiplicans,* respectively. Gradually, writers such as Pacioli (1494) and Licht (1500) omitted the word *numerus* in their descriptions of multiplication, leaving only *multiplicandus,* whence multiplicand evolved. In a similar but more gradual manner, the word multiplier replaced *numerus multiplicans.* Authors of incunabula arithmetics, as in the *Treviso* case, had no special name for the resulting product of a multiplication. *Numerus productus* was used by some later writers such as Clichtoveus (1503), eventually shortened to *product* (*produit,* Trenchant [1566]; *produtto,* Sfortunati [1534]).

The *Treviso* considers three methods of

multiplication and in this respect differs from many other early books where a larger variety of multiplication methods are expounded upon; for example, Pacioli (1494) in his work presents eight techniques of multiplication:[5]

1. *Per scachieri,* Venetian for tesselated. In Florence this method was called *bericuocolo,* the name of a pastry whose design bore a resemblance to the computational configuration.
2. *Castellucio,* Florentine for "little castle."
3. *A Traveletta* or *per colona,* by the table or column.
4. *Per quadrilatero,* by the *quadrilateral.*
5. *Per crocetta* or *casella,* by the cross or pigeonhole.
6. *Per gelosia* or *graticola,* method of cells.
7. *Per repiego,* method of decomposition of factors.
8. *A scapezza,* distributing, separating and multiplying by the parts, not the factors, of the multiplier.

Perhaps building upon his experience, the *Treviso* author makes an optimum selection of only three methods: *per colona; per crocetta* and *per scachieri,* in which to instruct his readers.

Multiplication by the Column

Almost all early writers of arithmetic texts began their considerations of the operation of multiplication with tables, both to instill the basic facts and to serve, in themselves, as a reference in performing simple multiplication exercises. In his construction of columnar tables, our author takes advantage of the identity property of multiplication by one, which he previously demonstrated in his consideration of numeration, and commutativity to effectively list only forty-four basic multiplication facts rather than one hundred. Then he lists five specific tables (i.e., multiplication by 12, 20, 24, 32 and 26) useful for conversions involving monetary and quantitative measures. Actually, the knowledge, construction, and use of such tables was of extreme importance to the trading professions of this time. A problem given in Johannes Widman's arithmetic of 1489 amply illustrates the need and use for such references:

> A man goes to a money-changer in Vienna with 30 pennies in Nuremberg currency. So he says to the money-changer, "Please change my 30 pennies and give me Vienna pounds for them as much as they are worth" and the money-changer does not know how much he should give the man in Viennese currency. Thus he goes to the money office, and they there advise the money-changer and say to him, "7 Vienna are worth 9 Linz, and 8 Linz are worth 11 Passau, and 12 Passau are worth 13 Vilshofen and 15 Vilshofen are worth 10 Regensburg and 8 Regensburg are worth 18 Neumarkt and 5 Neumarkt are worth 4 Nuremburg pennies." How many Viennese pennies do 30 Nuremberg pennies come to?[6]

Since many such tables concerned systems of Venetian measure or currency, they earned for themselves the descriptive reference *per Venetia,* literally, as done by the Venetians (Tartaglia, 1556). The *Treviso's* limited development of these supplementary tables was probably responsible, in part, for its lack of widespread popularity—it was just not complete enough a reference for commercial needs.

After enjoining the student to learn these tables by heart,[7] the author gives general directions on how to multiply "by the column" considering only multiplications that involve a single digit multiplier. Once again, the casting out of nines is suggested as a check of the correctness of the calculations performed. Some authors preferred a casting out of sevens. Reference is made to the "better" method, use of the inverse operation, division, as a check, but a fuller discussion of this method is delayed until division is formally considered.

Cross Multiplication

Crocetta, or as in the *Treviso, per croxetta simplice,* was a very common method of multiplication in old Italian arithmetics. Perhaps the best way to understand this algorism is by use of a diagram enumerating the distinct steps taken:

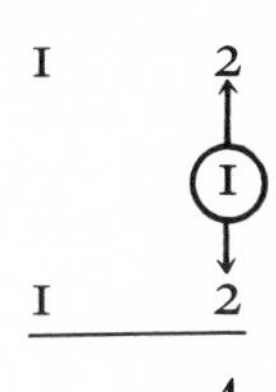

1. First multiply the unit terms together, writing down the unit result, and hold the ten's result, if there is one.

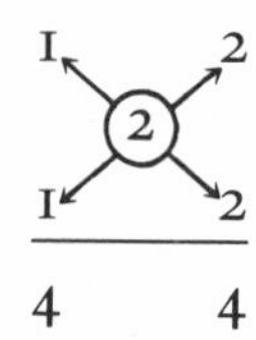

2. Multiply the units and tens terms together. Add to this the tens obtained in the units multiplication and write down the resulting number of tens. If any hundreds are obtained they are held for the next step.

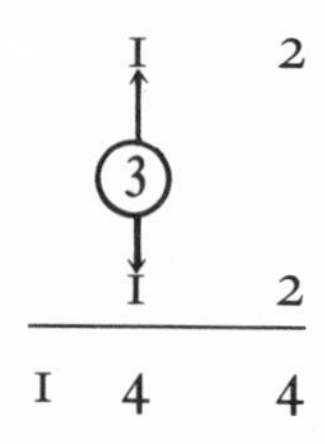

3. Multiply the tens terms together, add any tens carried over from the previous step and write down the result.

Some early works did have such illustrative diagrams. In accommodating these diagrams, printers often placed an *X* between the multiplicand and the multiplier:

$$\begin{array}{c} 3\ 2 \\ \times \\ \underline{2\ 5} \\ 800 \end{array}$$

This *X* clearly distinguished the problem as a multiplication exercise and the *X* evolved as the symbol for the operation. This process was awkward for factors with more than two digits but could be carried out as demonstrated in a problem from Pacioli's arithmetic (1494):

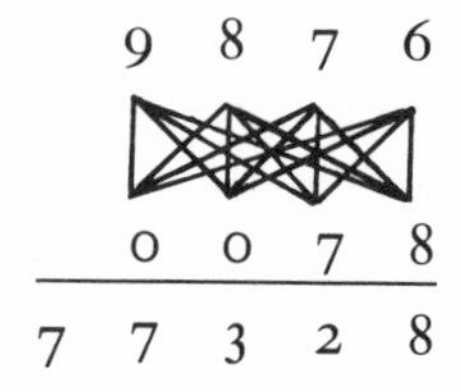

The cross method of multiplication can be traced back to Indian works of the twelfth century, *Lilavati* (c. 1150).

Chessboard Multiplication

The author uses the term chessboard multiplication, *moltiplicare per scachiero,* in the broadest sense to designate any multiplication algorithm based on the arrangement of numerical entries in square cells.[8] The initial example performed employs *per scachieri* technique; however, by this time the distinguishing grid was no longer used, but implied. This problem involves finding the product of 24 and 829 and the student reader is led through the steps to produce the written example:

```
      8 2 9
        2 4
  ---------
    3 3 1 6
  1 6 5 8
  ---------
  1 9 8 9 6
```

In an orthodox *scachieri* algorism the example would appear as:

			8	2	9
				2	4
	3	3	1	6	
1	6	5	8		
	1	9	8	9	6

The use of such cells ensures the preservation of proper place value during the multiplication process.

The variants of the chessboard method discussed on F21v and F22r are actually three different multiplication schemes according to Pacioli's system of classification. The first problem encountered on both folios is actually a *scachieri* multiplication with the multiplier written along the side of a slanted line which is intended to ensure the indexing of partial products to the left as place value increases.

```
       9 3 4                         5 6 7 8 9
    ----------                  --------------
     3 7 3 6 / 4                  2 2 7 1 5 6 / 4
       9 3 4 / 1                  1 7 0 3 6 7 / 3
   2 8 0 2 / 3                  1 1 3 5 7 8 / 2
    ----------                  5 6 7 8 9 / 1
   2 9 3 2 7 6          Sum     --------------
                                7 0 0 7 7 6 2 6
```

Many writers of arithmetics employed this scheme.

Although the form of the adjacent algorithm on F22r eventually became accepted as standard, this acceptance was long in coming. Smith has recorded chronologically the variants of the chessboard form as they appeared in various editions of Taglientes' *Libro dabaco* during the sixteenth century:[9]

```
   (1515)          (1520)              (1541)

    4 5 6           4 5 6               4 5 6
      2 3             2 3                 2 3
    -----           -----
  1 3 6 8       1 3 6 8               1 3 6     8
  9 1 2         9 1 2                 9 1 2
  ---------     -----
1 0 4 8 8       1 0 4 8 8           1 0 4 8     8
```

```
      (1547)              (1550)              (1561)

      4  5  6          4  5  6          4  5  6
      2  3                2  3             2  3
   ---------        ---------        ---------
   1  3  6  8       1  3  6  8       1  3  6  8
   9  1  2             9  1  2       9  1  2
------------     ------------     ---------------
1  0  4  8  8    1  0  4  8  8    1  0  4  8  8
```

```
        (1564)                    (1567)

        4  5  6                4  5  6
        2  3                   2  3
     ----------             ----------
     1  3  6  8          1  3  6  8
        9  1  2                9  1  2
     ------------      ---------------
     1  0  4  8  8      1  0  4  8  8
```

No doubt some of these forms may be due to typesetting errors, but still enough true variation exists to indicate that the standardization of algorithms had a complex history.

The second alternative chessboard technique considered is what was commonly known as *per quadrilatero* and can easily be seen to have evolved from *scachieri* techniques:

		9	3	4		
		3	7	3	6	4
			9	3	4	i
		2	8	0	2	3
2	9	3	2	7	6	

		5	6	7	8	9		
	2	2	7	i	5	6	4	
	i	7	0	3	6	7	3	6
	i	i	3	5	7	8	2	2
		5	6	7	8	9	i	6
Sum		7	0	0	7	7		

The multiplicand is written out along the top of a rectangle and the multiplier along the right side. Thus an array of cells is established whereby digits of the multiplicand designate columns of the array and digits of the multiplier designate rows. The column-designator digits are multiplied by the row-designator digits, resulting in a one- or two-digit product. The unit digit of these products is written in the corresponding cell and the tens digit carried to be added to the next row-column product. When all possible row-column multiplications are completed, the entries of the cells are added along diagonals from the upper left to the lower right, with the units digit of these sums written as an entry of the product and the tens entry carried to be added to the sum of the next diagonal column.

A ready variant of this method is the following *gelosia* or *graticola* technique, so named for its similarity to the contemporary lattice grillwork used over the windows of the high-born Italian ladies to protect them from public view. Following Byzantine custom, the wives and daughters of Venetian nobles were usually kept sequestered. Spying unobserved

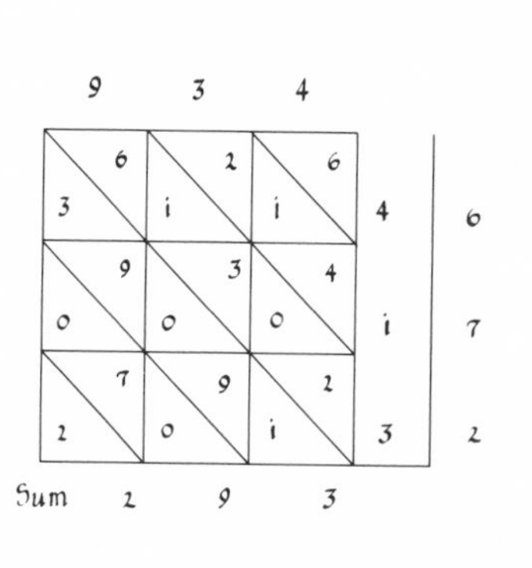

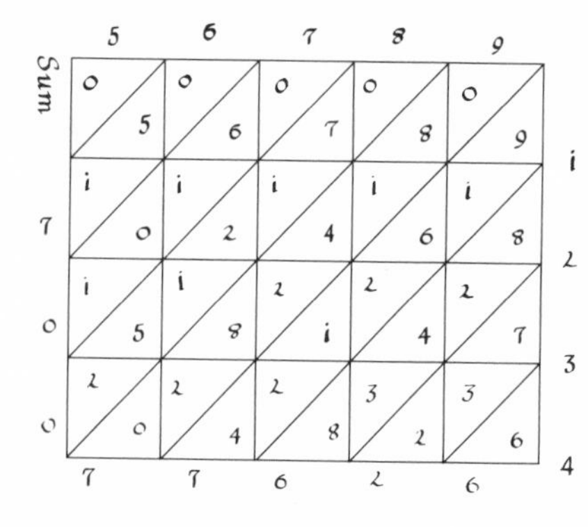

from such vantage points, these ladies often saw scenes that disturbed them or made them "jealous."

The computational technique employed is basically the same as that used in *per quadrilatero;* however, the cells are partitioned by diagonal lines so that when the product of a row entry and a column entry result in two digits the units digit is written below the diagonal line and the ten's digit above. In this manner, carrying becomes an integral part of the algorithm. When all partial products are obtained, the entries within diagonal columns are added as before and the resulting total product written along a side and base of the array.

This method is quite old. It probably originated in India, was known to be popular in Arab and Persian works, and was finally accepted into European arithmetics in the fourteenth century.[10] Due to its organizational efficiency and the ease it provides in multiplying any two multidigit numbers, it was quite popular as a computational scheme; however, it was difficult to print and read and thus fell out of favor. It is from the *gelosia* grid and principle that the computational device known as Napier's Bones (1617) evolved.[11]

✠ 5
Division

Division is described as the fourth operation, *quarto atto,* when actually within the scope of this work it is the fifth. Two names were used to indicate this mathematical operation, division and partition, and they were considered synonomous at this time. The *Treviso* actually speaks of partition, *partire,* the particular term that was popular in Italy and other Latin countries. Many writers such as Ghaligai (1521), Cataldi (1577), Ortega (1512), Huswert (1501), Stifel (1554), and Ramus (1555) also preferred this name. Authors of more theoretical books often used both forms (e.g., Pacioli [1594], Tartaglia [1556] and Clavius [1583]). Several other concepts of division were also employed in early arithmetic books, including one viewing it as an instance of repeated subtraction, a natural description of the physical operation of division as performed on an abacus.

The particular definition of the operation

itself, as presented in the *Treviso Arithmetic,* is the converse of the previous definition given for multiplication: quite simply, division is viewed as the inverse of multiplication—"division is the operation of finding, from two given numbers, a third number which is contained as many times in the greater number as unity is contain in the lesser number."[1] This definition is quite old and can be traced back to Maximus Planudes (c. 1340).

Early terminology used in explaining the division process is as varied as the definition of the process itself. Early Latin writers using simple descriptive terms spoke of the *numerus dividendus,* number to be divided, and *numerus divisor,* the dividing number. The contemporary Italian practice was to call the divisor "the parter," *il partitore,* and to avoid reference to the dividend; this convention is followed in the *Treviso.* Latin writers preferred *quotiens,* "the how many," for the quotient itself, although this term was subject to a vicissitude of alternatives, including "the product" (Frisius, 1540) and "the part," *la parte,* implying that the divisor is this part of the dividend, as employed in the *Treviso* work. The medieval Latin terms for the remainder were numerous: *residuus* in Grammateus (1518); *residius* in Scheubel (1545), and *residua* in George of Hungary (1498). The Treviso's author uses *lauanzo,* remainder.[3]

Division was considered the most difficult operation to teach and for the student to understand; as Pacioli (1494) noted, "if a man can divide well, everything else is easy, for all the rest is involved therein."[2] A knowledge of division implies a proficiency in the three other basic operations. Even by the time of the 17th century, when the Hindu-Arabic numeration system was fairly well established in Europe, the operation of division was still viewed with respect because of its difficulty:

> Division is esteemed one of the busiest operations of Arithmetick, and such as requireth a mynde not wandering, or setled uppon other matters.[4]

In large part, the ill repute of division resulted from its tedious and complex solution process as performed with counters. In referring to division carried out on the abacus, a medieval writer describes a computing situation involving "rules which the sweating abacists scarcely understand."[5] One such abacus-centered method of division was termed "iron division" because it was "so extraordinarily difficult that its hardness surpasses that of iron."[6]

Dura cosa e la partita—division is a difficult thing

The most straightforward algorithm for division, commonly known as short division, simply reverses the process for multiplication by the column or table, *per colona,* or, as described in other works, "by the rule" (Pacioli [1494]), and "in the head," "orally," and "to the right" (Tartaglia [1556]). The *Treviso's* instructions urge the students to begin with the highest place-value digit of the dividend and to see by the tables if the divisor goes into it; if not, they are to move one position to the right and repeat this inspection with the two-digit partial dividend. When division is possible, the partial quotient is written below and the remainder carried, in a manner similar to the transfer of tens in addition, to the next position and the process is repeated until no more integral divisions are possible. The remainder, if any, is written down separated from the quotient proper by use of a bar.

The divisor	2	7	6	2	4	0 The remainder
The quotient		3	8	1	2	

Multiplication of the quotient by the divisor is readily given as a method of ascertaining the correctness of an answer. The casting out of nines is also once again considered as a method of checking the solution.

This algorithm, intended mainly for problems involving single digit divisors, was not popular with students. In order to facilitate and promote its use, some Italian writers of the 16th century provided division tables in their books for reference.[7]

Perhaps the most impressive algorithm of basic computation used up through the 17th century was the galley, *galea* or *battello,* method of division, so named for the configuration of its digits, which reminded early writers of the sails of a ship. Illuminated manuscripts incorporated the outline of a galley about the computation, and Tartaglia mentions in his work that some Venetian teachers also required such illustrations from their arithmetic students[8] (see Figure 5.1). The name of the method appeared in various forms: *galea* in Pacioli (1494), and *galera* in Forestani (1603) and Tagliente (1515); however, in the vicinity of Venice, *battello* was used, and this is the

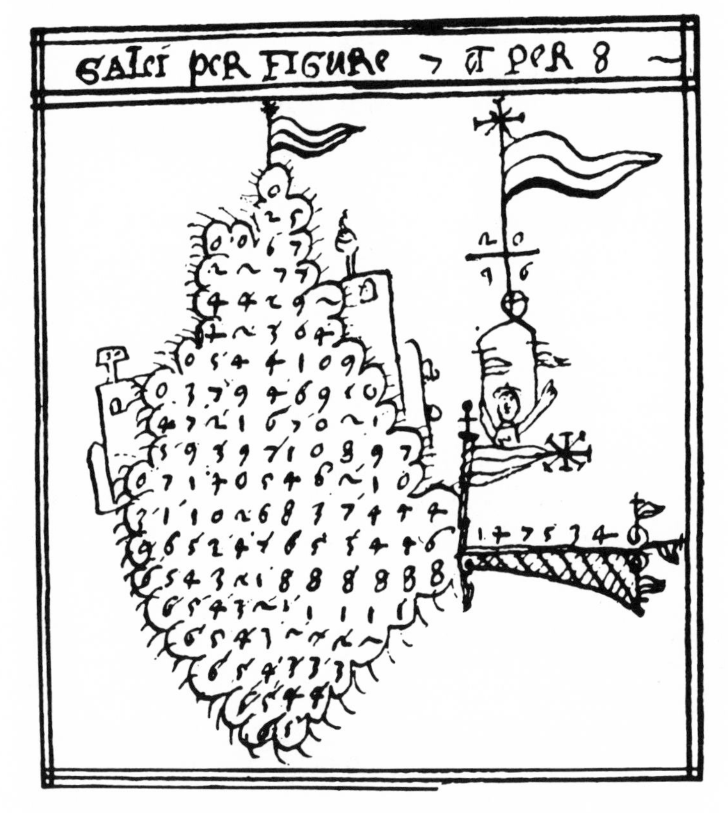

Figure 5.1 A sixteenth-century illustration of galley division done by a Venetian monk.

term employed in the *Treviso Arithmetic*. (It is interesting to speculate that if Venice had not been such a maritime power and the Venetians not so closely associated with ships, the algorithm might have borne a different name.[9] It seems, in the history of mathematics, that even the terms used are more frequently a product of the social, cultural, and economic environment rather than of the mathematics in question.) This method was also referred to as "scratch division" due to the necessity of crossing or scratching out various digits throughout the process. Maximum Planudes (c. 1340), in speaking of its early history, clarifies the name:

> [it is] very difficult to perform on paper, with ink, but it naturally lends itself to the sand abacus. The necessity for erasing certain numbers and writing others in their places gives rise to much confusion where ink is used, but on the sand table it is easy to erase numbers with the fingers and to write others in their places.[10]

This method can be traced to Eastern societies, Hindus employed it on their sand tables, and the counting-board techniques of Chinese arithmeticians closely approximated it.[11]

While there were many variations of the basic galley scheme, the algorithm presented in the *Arithmetic* is carried out as follows:

Consider the case of 1728 ÷ 12:

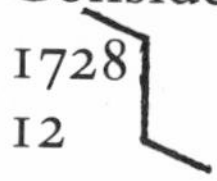

1728
12

1. The divisor is written directly below and at the left end of the dividend.

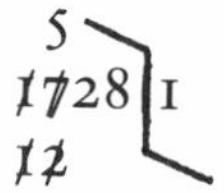

5
1728 | 1
12

2. The quotient will be placed to the right of the dividend and the remainder directly above the dividend. When a partial division is undertaken, the divisor is "scratched out" together with the digits it has been divided into.

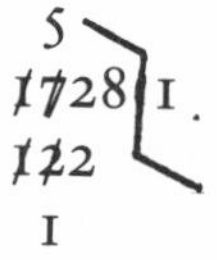

5
1728 | 1.
122
1

3. The divisor is advanced one place to the right after each division and written on adjacent lines, a process

spoken of by Sacrobosco (c.1230) as *anteriore* or *anteriorati,* and by Chuquet (1484) as *anteriorer.*

```
 54
1728 | 14
122
 1
```

4. A division is then undertaken with the uppermost digit and its diagonally adjacent companions serving as the dividend.

```
  1
 54
1728 | 144
1222
 11
```

5. This process is repeated until no more integral divisions are possible.

Occasionally, in texts, the illustration of this method appeared without cancellation marks—the printer lacked sufficient type. Due to the specific demands of mathematical typesetting, the printing of early arithmetic books was considered a challenge to the printer's ingenuity.

The prolonged historical popularity of the galley method of division was due in great part to its efficiency and resulting economy. The galley algorithm is actually more compact and uses fewer figures than the standard long-division method, and as a result requires less expenditure of paper to carry out. Since paper was still an expensive commodity at this time, the saving of paper was an important consideration in mathematical computations. An inspection of the same problem performed with the two different algorithms readily reveals the difference:

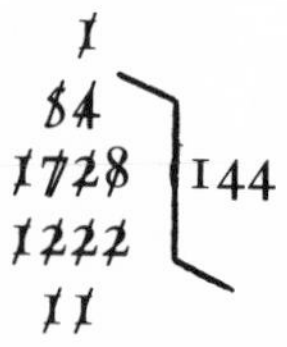

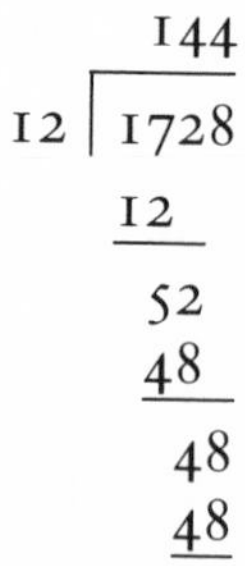

```
   1
  54
1728 | 144
1222
  11
```

```
        144
   12 | 1728
        12
         52
         48
          48
          48
```

Many early arithmetics openly advocated use of the galley method of division: Pacioli (1494) describes it as safer and more rapid than other methods and with "this galley" he proposes to sail the sea of arithmetic as he would the ocean itself; Hodder (1672) notes that he "will leave it to the censure of the most experienced to judge, whether this manner of dividing [galley] be not plain, lineal and to be wrought with fewer figures than any other which is commonly taught."[12]

Due to the perceived difficulty of the division process, the author of the *Treviso* is most specific and detailed in his instructions, leading the student readers through the technique with one-, two-, and finally three-digit, divisors. Furthermore, in considering only two methods of division, he once again exerts a judgment based on pedagogical and pragmatic experience.

Some Other Methods of Division

A few writers of the time presented more methods of division in their arithmetic books. For example, Pacioli gave four techniques: *a regola* or *a tavoletta,* "by the table"; *per repiego,* in parts; *a danda,* "by giving"; and *a galea* or *per galea,* as previously considered.[13] The present-day form of long division, although still evolving, was known at this time and advocated by many writers (e.g., Calandri [1491], Pacioli [1523], Tartaglia [1556], and Trenchant [1571]). In particular, Florentine arithmeticians were attracted to it, and termed it *"a danda."* The rationale for this term was explained by Cataneo (1546), who noted that during the division process, after each subtraction of partial products, another figure from the dividend is "given" to the remainder. This method first appeared in print in Calandri's 1491 arithmetic book[14] (see Figure 5.2). Throughout the 16th century, its acceptance and popularity increased and it received frequent mention by such Italian writers as Cataneo (1546) and Sfortunate (1544). Pagnini (1562) and Borghetti (1544) refer to it as "great division," whereas Pagani (1591) thought it "beautiful and pretty." Authors of arithmetics outside of Italy spoke of it as an "Italian method" and, indeed, it kept this association well into modern times.[15] By the 17th century, the algorithm of galley division had been replaced by the "danda method" as the conventional technique of performing division. But despite the acceptance of "downward" division as the primary algorithm for the operation, it would yet remain a long time before the quotient was to be written above the dividend. The mechanics of the galley technique had prevented the quotient from appearing above the dividend and, once positioned to the right of the dividend, it remained there by custom even

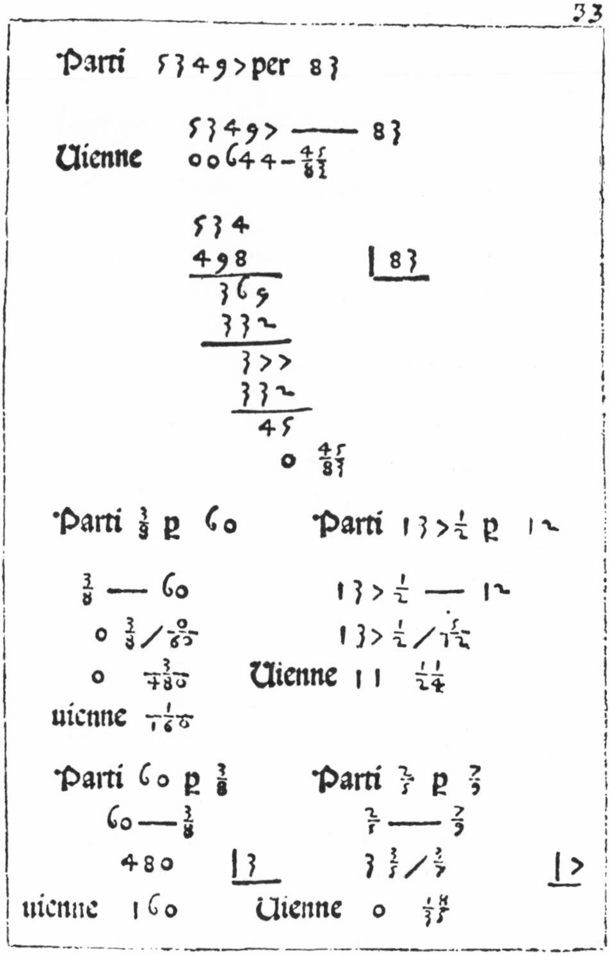

33

Parti 53497 per 83

53497 —— 83
Vienne 00644 — 45/83

534
498 | 83
369
332
377
332
45
0 45/83

Parti 3/8 p 60 — Parti 137 1/2 p 12

3/8 — 60
0 3/8 / 0/60
0 3/480
vienne 1/160

137 1/2 — 12
137 1/2 / 1 5/12
Vienne 11 11/24

Parti 60 p 3/8
60 — 3/8
480 | 3
vienne 160

Parti 2/5 p 7/9
2/5 — 7/9
3 2/5 / 7/9 | 7
Vienne 0 18/35

Figure 5.2 First printed example employing the modern form of long division (53, 497 ÷ 83), from Calandri's arithmetic of 1491.

though its parent algorithm had changed. It was the advent of the use of decimal fractions[16] and the need to preserve place value that finally relocated the quotient above the dividend.

A minor method of division also in use at

the time and worthy of mention is division *per repiego,* in parts, or *per regola,* by the rule. It is basically a division by factors; thus, instead of dividing 216 by 24, one would divide by 8 and then by 3—a technique easily undertaken with the use of tables. While this procedure may seem to be an inefficient expenditure of effort, it should be remembered that in the days when division was a "*dura cosa,*" such "shortcuts" were welcomed.

✠ 6
Applications: "What tools to use . . ."

Really, the discussion of the basic operations of mathematics in the *Treviso Arithmetic* are but a preparation for commercial problem solving. The relative importance placed upon this aspect of learning is emphasized by the fact that the remainder of the book, actually the majority of pages,[1] is devoted to considerations of various problem situations and their solutions. In the early arithmetics, the amount of space devoted to applications depended on both the nature of the work and the period in which it was published. Theoretical books, in general, contained fewer actual problems than practicae, and these usually reflected the interest and experience of their authors, which were not necessarily commercial in nature; thus, for example, a writer such as Champenois based his problems on military matters. Up until the middle of the sixteenth century, due to the expense of paper and the weak state of the art of printing, mathematics books were concise in their presentations, limiting the amount of space that could be spared for problem explanations. But, from the sixteenth century onward, the situation changed and larger selections of problems appeared in mathematics books.

As a commercial arithmetic, the *Treviso* concentrates on problems involving mercantile pursuits. In particular, its problem situations can be categorized as to the principles they encompass:

1. Use of the Rule of Three
2. Tare and Tret
3. Partnership
4. Barter
5. Alligation
6. Rule of Two
7. Pursuit
8. Calendar Reckoning

At this point in the students' studies, the author assumes a limited proficiency of the basic operations and offers the students problem situations that may further sharpen and strengthen their understanding of the techniques already discussed. Thus, there appears to be a dual purpose in introducing applications: first, to teach the rules and techniques of doing commercial mathematics, and second, to strengthen the basic computational skills just recently acquired by the students.

The Rule of Three

Perhaps no other mathematical technique was so esteemed in the late Middle Ages and early

Renaissance as the Rule of Three. The "Rule" was a simple proportion, although frequently ill-conceived, involving three quantities from which a fourth must be found. The problem was basic to all societies involved in trade and the exchange of commodities. Today such a problem would be considered trivial, but before the advent of mathematical or relational symbolism, when the dynamics of problems were presented rhetorically, the solving of such a problem posed considerable conceptual difficulties. The Rule of Three, as a mathematical technique, is quite old and can be traced back to problems in the Ahmes papyrus (c. 1650 B.C.) and the Chinese mathematical classic *Chiu chang suan shu* (c. 250).[2] Although the concept of proportion built on Greek traditions found its way into theoretical texts of this time, the Rule of Three obtained its exposure in commercial works and appears to have been an eastern importation.

The rule is found in the works of the Hindu mathematician Brahmagupta (628) under the name *trairasica,* the "three rule," and was cryptically explained as follows:

> In the Rule of Three, Argument, Fruit and Requisition are the names of the terms. The first and last terms must be similar. Requisition multiplied by Fruit and divided by Argument, is the Produce.[3]

When the rule appeared in Europe, the Hindu name was translated directly; however, the eastern terms for the parts of the proportion were not retained. Italian manuscripts of the fifteenth and sixteenth centuries usually had a section devoted to it, *De regola del tre.* In the *Treviso Arithmetic* this section is called *La regula de le tre cose,* the rule of three things.[4] Pacioli (1494) referred to it by both Italian and Latin names: *la regola del 3* and *regula trium rerum.* Similar names were employed by other Italian writers of the time. German authors copied the name used by the Italian merchants (e.g., Rudolff [1526], *Die Regel de tri*). Other names were also attached to it, attesting to its origins—Merchant's Rule—and to its importance—the Golden Rule.[5] This last title seems to have originated in the Germanic North, although Latin writers also preferred it. Petzensteiner, in his Bamberg arithmetic of 1483, uses this term with the rationale that the rule stands to other practices of arithmetic as gold compares with other metals in value; however, he dutifully credits its origin to Italy. From southern Europe, use of the rule spread to France and England where, in 1672, Hodder reiterates continental sentiment by noting:

> The Rule of Three is commonly called the Golden Rule, and indeed, it might be so termed: for as gold transcends all other Mettals, so doth this rule all others in Arithmetick.[6]

Its utility and versatility warranted admiration throughout the mathematics community of the early Renaissance. Clavius (1583) said it could not be sufficiently praised. Van der Schuere (1600) devoted a fifth of his book to the rule and its use; Cardinael (1659) assigned 40 percent of his tome to the rule. Perhaps the feelings of the computers of the time were best summed up by Baker (1568) who, in praise, stated:

> The rule of three is the cheefest, the moste profitable, and the moste excellente rule of all the rules of Arithmeticke. For all other rules have need of it, and it passeth all other.[7]

The rule was presented without a mathematical rationale, as commercial reckoning clerks were not interested in theoretical justifications of their techniques but rather in the quickness and accuracy of the results obtained. It was merely a procedure to be memorized and used. Authors of commercial arithmetic books took extensive care in establishing the manner in which the proportion was set up and solved.

Consider the following abstract form of a Rule of Three proportion: a is to b as c is to x. The *Treviso*'s author calls a "the divisor," *lo partitore;* b "the thing which has none similar to it," *la cosa che non ha somiglia,* and c "the thing which you wish to know," *la cosa che se vuol sapere*. The entities are set up in a row and the operations performed accordingly:

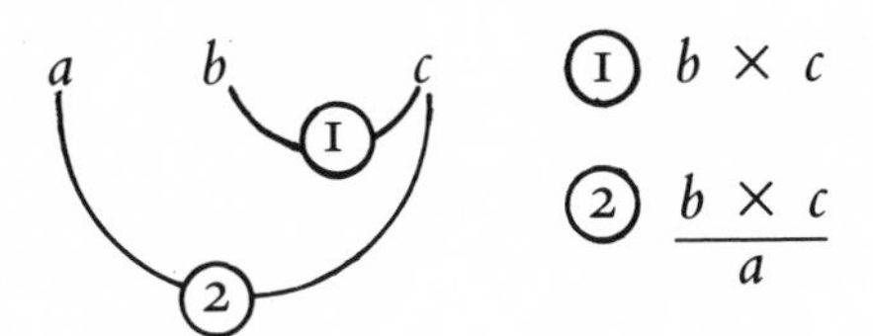

① $b \times c$

② $\dfrac{b \times c}{a}$

The resulting quotient will supply us with the unknown, in our case, x. This procedure is very similar to the modern notation $a: b:: c:x$ and, of course, gives the same result. Following an explanation of the rule, the author proposes fifteen problems graduated in difficulty from numbers without units attached (i.e., abstract numbers) to mixed fractions. Apparently to better prepare his students to work with fractions, the author begins by expressing whole numbers as fractions with a unit denominator:

$$\frac{8}{1} \times \frac{11}{1} \text{———} \frac{12}{1} \Rightarrow \frac{11 \times 12}{8} = \frac{132}{8} = 16\frac{1}{2}^{8}$$

It is interesting, and a bit disturbing, that no real instruction on fractions has preceded their rather extensive use in this section. It appears that a fraction is viewed as a variation of the division process and, thus, in the author's opinion, the practice already undertaken in learning division permits the use of fractions without further explanation. This was one prevailing concept of a fraction held at this time, the other being one or more equal parts of a whole. In the succeeding problems, practice is required in changing mixed numbers to improper fractions:

$$5\frac{3}{4} : 8\frac{1}{2} = 9 : x$$

$$\frac{23}{4} : \frac{17}{2} = 9 : x^{9}$$

and in denominating numbers into appropriate subunits, requiring a facility with monetary exchange:

$$1 \text{ ounce: } 4 \text{ lire, } 6 \text{ soldi} = 2\frac{1}{2} \text{ marks: } x$$

$$1 \text{ ounce: } 86 \text{ soldi} = \frac{40}{2} \text{ ounces: } x^{10}$$

Rather cumbersome fractions are introduced

into the computation, e.g., $\frac{3345332}{4320864}$, probably to test the student's stamina in problem solving.[11] In problems where the divisors are multiples of 100 and 1000, the author introduces the computational shortcut of moving the required number of decimal places in the dividend and performing the division by the digits remaining in the divisor.[12]

In such a division, the resulting decimal fraction component of the quotient is denoted by the use of a stroke and half bracket; for example, on F37r we encounter the problem where 2016 is to be divided by 100 and the result is given as 20 $\lfloor\underline{16}$. This notation reflects an appreciation of the mechanics of decimal notation and was employed by other authors of this period. Pietro Borgi (1484) used it in his commercial arithmetic and Francesco Pellos published a practica at Turen in 1492, in which he actually used a dot to separate the integer part of a number from its fractional component. Unfortunately for the history of mathematics, Pellos did not pursue or extend the development of his notational innovation.

After a discussion of the Rule of Three and its applications involving direct proportion, the *Treviso* author then briefly considers the Rule of Three involving inverse proportions.[13] This procedure went by many names: the inverse, converse or everse Rule of Three, and in England it was called the Backer rule. Relying on modern expression and symbolism to understand the *Treviso*'s instructions, if given a problem of the form 'a is inversely proportional to b as c is inversely proportional to x', this would be symbolized as:

$$a : \frac{1}{b} = c : \frac{1}{x} \text{ or}$$

$$\frac{\frac{a}{1}}{\frac{1}{b}} = \frac{\frac{c}{1}}{\frac{1}{x}} \text{ and solved}$$

$$\frac{a \times b}{c} = x$$

The author asks his readers to set up the Rule of Three as in previous problems, only to interchange the first and last terms and proceed with the agreed-upon technique. Thus the solution is found:

$$\frac{c}{1} \times \frac{b}{1} \text{ —— } \frac{a}{1} \Rightarrow \frac{b \times a}{c} = x$$

Early writers did not invert either ratio; they left their terms in the general form of a direct proportion, but merely changed their solution rule. Hylles, in 1600, explained the process followed by most arithmeticians of the time:

> The Golde rule backward or converst, Placeth the termes as doth the rule direct: But then it foldes the first two termes rehearst, Dividing the product got by that effect. Not by the first, but onely by the third, So is the product the fourth at a word.[14]

Thus, they would use the same format as the Rule of Three, that is, without the first and last terms interchanged, but would reverse the operations so that the result is the same as obtained above. The *Treviso*'s technique emphasizes that a different mathematical situation is being considered.

In this period of history, the solution of problems by means of direct or indirect proportion was considered a powerful mathematical technique. Many problems that today would be solved by the methods of elementary algebra were manipulated and interpreted so that a technique employing proportions could be adopted to obtain a solution. While the Rule of Three was the most popularly-employed solution scheme involving proportions, there also existed a Rule of Two which the *Treviso*'s author also demonstrates, and which will be examined and discussed later in this chapter. Authors of other commercial arithmetics sometimes considered the Rule of Five, *regole 5 cose:*

> If a problem has been given in which 5 things are proposed, we must multiply the thing that we want to know by the two which are not similar and divide by the two similar.[15]

They also provided problems to demonstrate the technique:

> For example, 4 braccia of cloth are worth 5 florins and 6 florins are worth 4 lire. What are 124 braccia of cloth worth?[16]

In addition to these, they sometimes used a Rule of Seven. Merely by adding additional ratios, a compound proportion, or "chain," of any length could be established.

In reading the author's instructions of just how to apply the Rule of Three, one is struck by the repetition of the word *"cosa"*—thing (i.e., "the thing which has none similar to it," "the thing which is mentioned once," "the thing which you wish to know," etc.).[17] The repetition of this word and its associated phrases is monotonous and begs some decisive action to alleviate the situation. A modern reader of these passages readily realizes that the "things" or unknowns described could be listed in a symbolic form as variables and, indeed, this practice was adopted by some Italian writers at this time. Pacioli in his *Summa* (1494) moves from a purely rhetorical algebra as employed in the *Treviso Arithmetic* to one syncopated by the use of abbreviations for frequently referred-to terms and operations.[18] In the *Summa, cosa* is abbreviated by *co.* Early Italian writers called their algebra, or study of unknowns, the *"Regola de la Cosa,"* Germans adopted the term *"Die Coss,"* and the English eventually spoke of the *"cossike art."*[19]

Tare and Tret

Although this topic is extinct in modern-day arithmetic books, it was popular and important up until a century ago when the transportation of merchandise still involved cumbersome containers, casks, barrels, etc., and was time-consuming.[20] Tare is the determination of the weight of the cargo with a deduction made for the weight of the containers that bore it; the term "tare" was frequently used for the actual allowance provided (for example, 4 pounds per hundredweight). The word "tare" evolved from the Arabic *traha,* that which is thrown away. In the Italian of the fifteenth century, *tara* was the word used for tare, and this is the term employed in the *Treviso Arithmetic.* Tret comes from the Italian *tratto,* allowance for transportation, and actually was an allowance, usually in money, made to a buyer of certain goods for waste, damage, or deterioration of his merchandise incurred during transit. For example, brazilwood was bought in Alexandria and measured by the *cantare fulfuli,* a measure usually associated with the purchase of pepper. In transporting the wood to Italy six cantari were given for every five purchased.[21].

The problems used as illustrations in this section are rather straightforward but some bear certain features in their presentation or manner of solution which are unique or of historical relevance. In the first problem given, the reader is required to find the total tare allowed on 4562 pounds at 4 pounds per hundredweight.[22] A student solving such a problem today would find out how many hundredweight were in 4562 pounds and multiply by the given allowance, 4:

$$100\overline{)4562}^{\;45} \qquad \frac{62}{100} \qquad \begin{array}{r} 45.62 \\ \times\ 4 \\ \hline 182.48 \end{array}$$

Our author, however, reverses this order of solution by first multiplying 4562 by 4, obtaining 18248, and then divides this result by 100, 182 (48 = 182.48.

The phrase "by the hundred," or "hundredweight," in Italian was *per cento* or *per cent.* Italian merchants of the fifteenth century made such use of the term and for convenience abbreviated the expression to p c°, and finally c°, whence came the symbol % that is used to express percent today. The mathematical concept of percent, however, was already in use among Italians at an earlier period. Account books dating from 1307 have profits recorded as a percent,[23] that is, by considering the ratio of profits to equity, $\frac{10,670}{41,365}$.

Partnership

The history of commerce and business has passed through three relative phases of activity. In ancient times, when societal interaction was limited, trade could be carried out by a single individual, a traveling merchant, serving a local or regional area. As the accumulation of wealth and the needs of society increased, the demands of traders, both financial and physical, required a broader supporting base than that supplied by an individual; the result was the institution of partnership. The business arrangement of partnership can be traced back to ancient Babylonia (2000 B.C.). Finally, the ability of partnership arrangements to finance and carry out business transactions was superseded by that of the corporation, which also appeared fairly early in the course of human history, being known in Roman times as *societates publicanorum*.

Perhaps well into modern times, the most pervasive form of mercantile institution has been the partnership. In the flourishing commercial climate of the Venetian Republic, partnership became an important aspect of business life and many types of this association evolved; for example, under an agreement called the *colleganza,* a merchant would contribute his skill and labor and his partner would be the financial backer for a commercial endeavor. In return for his services the merchant would keep $\frac{1}{4}$ of the profits and the financial partner would receive the rest.[24] Most partnerships, however, were on a financial basis and preferably undertaken between family members, albeit an extended family, or between individual traders. Two main reasons prompted the use of partnerships in business transactions: first, business investments tied up large sums of money for long periods of time (to purchase, trade, and receive returns on a cargo of pepper might take several years; few individuals could afford to have such an outlay of money for so long a period); and secondly, the merchants of Christian Venice and its surroundings were forbidden by Church law to promote usury. The Venetian interest rate for lending money at this time was about 20%; merchants could not lend excess capital on the open market and escape the condemnation of the Church, so the use of partnership avoided such encounters and yet provided equally attractive financial returns.

A compact of partnership required an equitable distribution of profits, and thus mathematics of partnership became an important aspect of business life. Most medieval writers of arithmetics considered partnership an important topic and sections on it can be found in the works of many authors including Leonardo of Pisa (1202) and Johannaes Hispalensis (c. 1140).[25]

The mathematics of partnership is found in

the Latin books under the title *Regula de societate* (Huswert [1501], *Regula de societate mercatorum et lucro*); Cardan (1539) uses the simple name *De societatibus*. Italian writers considered the computational processes under the term *compagnie,* as in Feliciano (1526), *Dele compagnie*. The *Treviso Arithmetic* merely speaks of "problems in partnership," *rarone di compagnia*. The Italian word *compagnia* passed directly into French, and finally into English as "company." Generally, English authors favored the name Rule of Fellowship, thus Recorde (1542) refers to "The Rule of Fellowshyppe or Company."

Arithmetic explanations of this rule were presented in two ways: without a consideration of the duration of the investment, it being assumed that all partners invested their money for the same period of time, and with a consideration of time. The problems presented in the *Treviso Arithmetic* represent both situations. A sufficient understanding of the nature of computation involving partnership can be arrived at by examining the workings of one problem in detail. The second problem given has Sebastiano and Jacomo enter into a partnership. Sebastiano puts up 350 ducats for a period of two years, and Jacomo invests 500 ducats, 14 grossi, for a period of 18 months; how should their profit of 622 ducats be divided?[26]

1. The amount invested is reduced to a common unit, grossi:

   Sebastiano, 350 ducats = 8400 grossi

   Jacomo, 500 ducats, 14 grossi = 12014 grossi

2. Each share is multiplied by the time invested in months:

Sebastiano, $8400 \times 24 = 201600$
Jacomo, $12014 \times 18 = 216252$

3. The shares in terms of grossi times months are added:

$$\begin{array}{r} 201600 \\ \underline{216252} \\ 417852 \end{array}$$

4. The Rule of Three is applied to determine Sebastiano's share of the profit:

$$\frac{417852}{1} \times \frac{622}{1} \text{ —— } \frac{201600}{1}$$

$$\frac{201600 \times 622}{417852} = 300.95 \text{ ducats}$$

300.95 ducats = 300 ducats, 2.28 grossi, since 1 grossi = 32 piccoli.

.28 grossi = 8.96 piccoli, or as the answer is expressed in the given solution, 300 ducats, 2 grossi

$8\frac{327456}{417852}$ piccoli.

5. Step 4 is repeated to determine Jacomo's share of the profit, which is found to be 321 ducats, 21 grossi 13 $\frac{90396}{417852}$ piccoli.

6. As a proof of the computation, the shares are added to see if their total equals the realized profit of 622 ducats.

300 d	$\frac{1}{2}$ g	$\frac{1}{8}$	$\frac{327456}{417852}$ P
321 d	21 g	13	$\frac{90396}{417852}$ P
622 d	24 (crossed out)	22 (crossed out)	$\frac{417852}{417852}$ (crossed out)

Barter

Barter, *de baratti,* the direct exchanging of a good for another good, was still a common mercantile practice in early Renaissance Europe. A contemporary work roughly defines the "Rule of Barter," "to keep oneself from being deceived or tricked and from deceiving or tricking others," and a later (1579) definition sets a worldly tone:

> Bartering is nothing but giving away one piece of merchandise for another in hope of coming out in a better position.[27]

By this time, the art of bartering had progressed to the stage where the exchange was not based merely on an object for an object, but rather set around a standard to which the value of the two objects could be reduced; this standard in most cases was a hard currency such as lire, ducats, etc. In bartering, the price was usually set higher than if the exchange was to be made on a money basis. This practice is demonstrated in the second problem of this section, where the cloth is stated to be worth 22 soldi a yard yet will be bartered at 27 soldi per yard. The reasons for preferring barter to direct monetary payment were twofold: there was a shortage of currency (not until the discovery of gold in the New World and its import into Europe did a large-scale money economy function), and the existence of trade fairs. Much of the high-volume trading done at this time was accomplished at inter-regional fairs where merchants from many countries would come together to exchange goods. In the absence of a trusted "international" currency, it was easier and safer to barter goods.

German books termed the computation involving barter *Stick Rechnung,* reckoning of mutual exchange (Heer [1617]). French writers used the word *troquer,* which passed into English as "truck," which came to mean the objects themselves that were bartered (e.g., garden truck). Later this term came to mean the vehicle used to carry the objects of barter, and remains with us today in this context.

Most commercial arithmetics had a section devoted to the topic of barter and some were quite large; for example, Baker (1568) devotes 15 pages of his book to this topic, "The Xl chapter treateth of the Rules of Barter: that is to change ware for ware," but the *Treviso*'s consideration of barter is brief. Only three problems involving barter are presented.[28] Actually, the mathematics involved is neither new nor complex; only the problem situations warrant attention, therefore it would seem that perhaps brevity of presentation was appropriate.

Alligation

Alligation as a topic of mathematical importance had first made its appearance in the arithmetic books of the fifteenth century in connection with metallurgy. The word has its origins in the Latin *ad,* to, and *ligare,* bind, and literally means "to mix." Alloy is derived from the same root. Dutch writers called it *menginghe,* mixtures; for example, Vander Schuere (1600) has a section in his book devoted to *Allegatio Menginghe.* As metallurgy, stimulated by the demands of bell and cannon casting and minting, began to be recognized as a science, consideration of its quantification techniques moved from the manuals of alchemists to those of the reckoning master. The alligation problems of the *Treviso Arithmetic* all deal with the processing of precious metal. Mints were common in the ancient and medieval world; city-states, of which Italy claimed 20 at this time, had their own, and even families through inherited rights could coin money. In the fifth century Sicily alone boasted the possession of over 50 mints. Without central controls over coinage, standards of minting varied greatly; thus, a knowledge of the mixture of precious metals was essential for the business world as the value of money was not determined by its face value, but rather its content of precious metal. Also at this period of European history, the mining industry was experiencing a revival. Gold and silver from new discoveries in Germany and Hungary found their way to Italy for refining and entry into the commercial world. Throughout the sixteenth century, chapters on alligation in arithmetic books were particularly intended for the German mintmasters, *Müntzmeister,* or the Italian goldsmiths, *"Del consolare dell'oro e dell'argento."* Later, the scope of the topic was expanded to include problems of mixtures in general. For example, Recorde (c. 1542) notes:

> it hath great use in composition of medicines, and also in myxtures of metalles, and some use it hath in myxtures of wines, but I wshe it were lesse used therin than it is now a dais.[29]

Thus by 1568, one finds 48 pages of Baker's book devoted to various applications of alligation.

The *Treviso*'s alligation problems merely illustrate another instance for the repeated use of the Rule of Three, as in the first problem.

A merchant has 46 marks, 7 ounces of silver in which he knows there is $7\frac{1}{4}$ oz. of pure silver per mark. He wishes to reduce the purity of his silver to less than half, down to $3\frac{1}{2}$ oz. of fine silver per mark. The question asks "how much brass must be added to accomplish this?"[30]

1. The total purity of the silver is determined: As 1 mark = 8 oz., therefore, 46 marks, 7 oz. = 375 oz., and the proportion is established.

$$1 : \frac{29}{4} = 351 : x$$

Using the Rule of Three:

$$\frac{8}{1} \times \frac{29}{4} \text{———} \frac{375}{1}$$

it is found that x = 339,843 oz. or 42 marks 3 oz., 3 qr., $13\frac{1}{2}$ k.

2. Given the amount of pure silver and a measure of the alloying factor or dilution required, a second proportion can be set up:

$$42 \text{ marks, } 3 \text{ oz., } 3 \text{ qr., } 13\frac{1}{2}\text{ k} = \frac{97875}{2}\text{ kauts}$$

$$\frac{7}{2} : 1 = \frac{97875}{2} : x$$

by the Rule of Three:

$$\frac{7}{2} \times \frac{1}{1} \text{———} \frac{97875}{2}$$

$$x = \frac{97875}{7} = 13977.857 \text{ kaut}$$

$$13977.857 \text{ kaut} = 97 \text{ marks, } 3 \text{ qr., } 5\frac{1}{7}\text{ k}$$

3. The amount of brass to be added is found by subtracting the amount of silver in hand from that to be alloyed:

97 M	0 oz.	3 qr.	$5\frac{1}{7}$ k
42 M	3 oz.	3 qr.	$13\frac{1}{2}$ k
54 M	4 oz.	3 qr.	$27\frac{8}{14}$ k

The Rule of Two

The Rule of Two, *el regula de le do cose,* involved the dividing of the product of two given numbers by their sum in order to solve a particular problem. The statement of this rule is deceptively simple and conceals some subtle mathematics that can best be appreciated by solving the given courier problem by modern methods.

We are given that one courier embarks to Rome from Venice and intends to reach there in 9 days. At the same time, a messenger from Rome is traveling to Venice and will reach his destination in 7 days. We are further told that the distance between the cities is 250 miles, and asked in how many days these travelers will meet.[31]

The Rome to Venice courier is traveling at a rate of

$$\frac{250}{7} \text{ m/day}$$

The Venice to Rome courier is traveling at a rate of

$$\frac{250}{9} \text{ m/day}$$

Since they will meet in a fixed number of days, *d*:

$$\frac{250}{7} d + \frac{250}{9} d = 250$$

$$250\, d \left(\frac{1}{7} + \frac{1}{9}\right) = 250$$

$$d\left(\frac{1}{7}+\frac{1}{9}\right) = 1 \text{ or } d\left(\frac{9}{9}\left[\frac{1}{7}\right]+\frac{1}{9}\right) = 1$$

$$\frac{1}{9}d\left(\frac{9}{7}+1\right) = 1 \Rightarrow d\left(\frac{9}{7}+1\right) = 9$$

$$d\left(\frac{9+7}{7}\right) = 9 \Rightarrow d = \frac{9 \times 7}{9+7}$$

this is the result provided by the *Treviso*'s author.

Courier-type problems were popular in old mathematics books and appeared in many variations, including couriers starting at different times or traveling over geometrically prescribed routes. An example in the Chinese work *Chiu chang suan shu* (250 B.C.) has the travelers progressing along the perimeter of a right triangle.[32] The first known writer to give a listing of variants of the basic courier problem was Pacioli (1494). In Petzensteiner's *Bamberg Arithmetic* (1483), the second arithmetic text to be published in German, there is a chapter called *Von Wandern,* in which is related the problem of two young men setting off at the same time for Rome. The first travels six miles per day, and the second progresses one mile the first day, two miles the second day, etc. It is required to find when the second traveler will overtake the first. Calandri (1491) offered a variant involving sailing ships, the first going from Pisa to Genoa in five days, and the second sailing from Genoa to Pisa in two. They start simultaneously from their respective ports and it is required to know when they will meet. Köbel (1514), in a chapter entitled *Von Wandern über Landt,* presented the situation concerning two citizens of Oppenheim, Heinrich and Contz von Treber, who set out from their home for Rome. Heinrich, being elderly, could travel only ten miles a day, while Contz von Treber was young and progressed thirteen miles per day. Heinrich began his journey nine days ahead of Contz, and it is required to know how long it will take Contz to pass Heinrich.

In the majority of courier problems, Rome is a destination or a starting point for a journey. This fact could either testify to the Italian origins of such problems, or else affirm the importance of Rome as a focus of political and religious activities in Europe at this period of history.

At this point in the text, the author introduces "the rule of two terms which fall and come together," *el riegula de le do cose che se cazano po se cõzongeno;* this involved problem solutions obtained by dividing a specific product by the difference of two numbers.[33] The first example given is the "famous" pursuit problem involving a hound and a hare, which was a standard exercise in European arithmetic books for centuries. Its European debut was in Alcuin of York's *Propositiones ad acuendos juvenes* (c.775), a book of puzzles presented to Charlemagne.[34] Books and manuscripts contemporary with the Treviso work also contain the problem (e.g., Pretzensteiner [1483], Calandri [1491], and Benedetto da Firenze [c. 1460]). Use of modern techniques, once again, reveals the validity of the solution procedure.

A hare is 150 paces ahead of a hound,

which pursues him. The hare covers 6 paces while the hound covers 10. In how many paces will the hound overtake the hare?[35]

1. Let t represent the time period in which the hare covers 6 paces and the hound 10, then:
   distance hare traveled $= 150 + 6t$
   distance hound traveled $= 10t$
2. Since they will have traveled the same distance when they meet:
   $150 + 6t = 10t$
   $150 = 4t$
   $37.5 = t$
3. The distance the hound will have traveled is:
   $10t = (37.5)(10) = 375$ paces

The following problem involves a purse of ducats. Once again, when the problem is solved by modern algebraic methods with the introduction of x for the unknown, its solution method is found to coincide with the author's directions:

A man finds a purse with an undetermined amount of ducats in it. He spends $\frac{1}{4}$, $\frac{1}{5}$, and $\frac{1}{6}$ of the amount and 9 ducats remain. It is required to find out how much money was in the purse.[36]

1. Allowing x to be the money, the problem is then denoted by the equation:

$$\frac{x}{4} + \frac{x}{5} + \frac{x}{6} + 9 = x$$

2. The least common denominator for the given fractions is 60, therefore:

$$\frac{15x + 12x + 10x}{60} + 9 = x, \text{ or}$$

$$\frac{37x}{60} + 9 = x$$

3. $9 = x - \frac{37x}{60} \qquad 9 = \frac{(60-37)}{60}x$

4. Therefore:

$(60)(9) = (60-37)\,x$ or $540 = 23x$
$x = 23.478$ ducats

The rule given in the book corresponds to step 4 in our solution process and the answer is found to be $23\frac{11}{23}$ ducats.

The remaining problems in this section are classics and have been perplexing schoolboys for centuries. For example, the problem concerning men building a house is a variant of an even more ancient and traditional problem, that of pipes filling a cistern or fountain, common sights in the Mediterranean world. This problem has been attributed to Heron of Alexandria (c. 75), and is known to have appeared in Diophantus' works (c. 275). By the time of the sixteenth century, it had become a stock problem and was adopted into mathematical books in rather absurd and colorful situations; for example, Cataneo (1546) gives a problem of wild animals devouring sheep, and Frisius (1540) poses the problem of a husband and wife drinking wine.

The Golden Number and the Full Moon

Procedures for determining the date of Easter and the phases of the moon are, by contemporary standards, strange information to find in a book ostensibly devoted to commercial arithmetic. But, of course, one cannot judge the contents of the *Treviso Arithmetic* by contemporary standards. For hundreds of years, mathematics and astrology were, in the minds of many, closely associated. The people of the Middle Ages and Early Renaissance were strongly influenced by this magical art that could seemingly associate the unknown, their well-being and destinies, with the known, the movement of heavenly bodies. Mathematicians were sought out for their knowledge and ability to perform astrological calculations. As a result of these demands, reckoning masters added this service to their repertoires. Even such accomplished mathematicians as Cardan and Kepler supplemented their incomes by casting horoscopes.

Merchants in their business ventures and travels continuously tempted the "Fates"; therefore, it is not strange to learn of their involvement with astrology. Advice such as "If the calends of January fall on Sunday . . . grain will be neither cheap nor expensive. If they fall on Monday . . . there will be plenty of grain . . ." influenced grain transactions.[37] Travel, an integral aspect of a merchant's life, was a hazardous endeavor and warranted astrological counsel.

> When the moon is in Gemini, it is good to get out of a port if the voyage had

> already begun . . . but if it has not, you should not sail, for it is a token of wind.
> Libra: it is good to begin all travel by ship or on land, to buy and to sell, for this is a token of air.
> Virgo: do travel by sea and land, buy, sell, begin anything, for this is a token of land.[38]

Some commercial reckoning books and manuscripts contained sections devoted to astrology, and the subject even found its way into the manuals of mercantile knowledge kept by certain trade houses.[39] As the scientific spirit emerged in Italian society, astrology disappeared from commercial arithmetics. Referring to astrology, Francesco di Carlo de Macigni advised in his 1457 *Libretto all praticha della merchatantia:*

> one who wants to be a merchant has no need to learn how to measure the stars and planets, the course of the planets, or the movement of the heavens.[40]

Pacioli, in his writings on the business world of the fifteenth century, also dismisses astrology. By including a brief section on the "Golden Number", *lo aureo numero,* and the phases of the moon, the *Treviso Arithmetic*'s author reflects both this new scientific spirit and merchants' continued concern with understanding and using forces beyond their immediate control.[41] At least two reasons can be given for a merchant's involvement with calendrical reckoning: market and trade conditions are frequently seasonal in nature, and the existence of civil holidays and Church feast days often prescribed constraints on the use of labor and money.

The calendar of Europe in the early Christian era and well into the Middle Ages was, in a sense, two calendars superimposed upon one another. One calendar marked the chronological passage of time, the other noted Christian feast days. The latter strongly affected the activities of daily life, as it exempted certain days from the pursuits of manual labor and financial transaction. It was important for the merchant-businessman, as well as the clergy, to know the specific dates of Christian holidays. Theologically speaking, the primary feast day of the Christian Church is Easter, which marks the resurrection of Christ from the dead and affirms His Divine association. Around the date of Easter is established a network of Christian holy days:

Septuagesima Sunday	9 weeks before Easter
Sexagesima Sunday	8 weeks before Easter
Quinquagesima Sunday	7 weeks before Easter
Quadragesima Sunday	6 weeks before Easter
Shrove Tuesday	Eve of Lent (period of Christian fasting and penance)
Ash Wednesday	Beginnings of Lent
Lent	40 days
Palm Sunday	End of Lent and beginning of Holy Week
Good Friday	Friday before Easter
Easter Sunday	
Rogation Sunday	5 weeks after Easter
Ascension Day	40 days after Easter
Whit Sunday	7 weeks after Easter
Trinity Sunday	8 weeks after Easter

From this list, it can be appreciated that Easter itself directly dominates about seventeen weeks of the Church calendar; the date of Easter becomes a critical factor in organizing one's life and business dealing in a devout Christian state.

Eastern Christians built directly upon the traditions of their Jewish precursors: they observed the Passover correlating Christ's last days in this Jewish context. As a result of this practice, their date for Easter was determined by the Jewish calendar, which is lunar in its conception.[42] Under Hebrew custom, the Passover begins at sunset of the fourteenth day of the lunar month, and as Easter was a part of Passover (for eastern Christians), its date was determined within the context of the month. Western Christians, following a solar-based Julian Calendar, insisted on celebrating Easter in the context of a week, Holy Week—the death of Christ on a Friday and His resurrection on the following Sunday. The two chronologies could not be reconciled, and a schism existed between Eastern and Western Christendom. Partly to solve this quarrel, Constantine the Great called the Council of Nicaea in 325 A.D. At this Council, advocates for the Western-oriented 'week-based' Easter won their cause. Henceforth, Easter was to be celebrated everywhere in the Christian world on the same day, a Sunday.[43] More specifically, the Council set the date of Easter within the Julian Calendar, but still included reference to the moon's position. Easter was declared to be the first Sunday after the full moon following the vernal equinox, March 21, and if the full moon happened upon a Sunday, Easter day would be the following Sunday.

The question of determining the occurrence of the full moon was referred to the Bishop of Alexandria whose city, at this time, served as the center of Western astronomical activity. Since the appearance of the moon is a cyclic phenomenon, a period for its determination had to be decided upon. The use of the Metonic cycle of 19 years, a period within which the full moon would reappear on the same day, was officially sanctioned and henceforth, using this information, the date of Easter could be predicted.[44] To accomplish this feat, one must know when the vernal equinox takes place, the occurrence of the following full moon, and a method of compensating for errors in the Metonic cycle. The astronomical problems involved in this computation were settled by the year 525 and tables were compiled by Dionysius Isidore and Bede to assist in the task of reckoning the date of Easter.[45] To be able to use such tables, one had to be familiar with the Golden Number. From the time of Charlemagne, every church official was required to understand the use of the Golden Number in calendric reckoning.

The Golden Number is found by adding 1 to the number of the year in question and dividing by 19—the remainder is the Golden Number. Instructions given in the *Treviso* differ slightly from this procedure, requiring the 1 to be added after the division by 19 was undertaken; however, the results of the two procedures are identical. By the time of the printing of the *Treviso Arithmetic,* Golden Number interpolation was incorporated into calendars themselves and thus, by knowing the Golden Number, one could determine the occurrence of the new moon for any given month.

In Germany, Gutenberg published calendars and almanacs from 1448 onward and Regiomontanus offered his readers calendars that, besides listing Christian feast days, offered forecasts concerning planetary constellations, eclipses, and other astrologically important happenings (see Figure 6.1). An extant German calendar has been found listing the most important days for the years 1478—1496 (see Figure 6.2). Its interpretation requires the use of the Golden Number. As indicated by the *Treviso*'s reckoning process, the calendar indicates that the Golden Number for the year 1478 is 16.

The author of the *Treviso Arithmetic* states the period between moons as 29 days, 12 hours, 793 points (*puncti*). Although the chronological measuring unit "point" is unknown today, we are also told that 1 hour = 1080 points,[46] from which we can gauge a point at 1.33 seconds, or the fact that 18 points = 1 minute. If this period is then compared against the modern calculations for the period of a new moon; 29 days, 12 hours, 44 minutes and 2.8 seconds, the error of the Renaissance calculation is almost negligible.

Following in the last section of the book is a listing of conversion rules for selected weights and measures. They were apparently appended as an afterthought.

Calender des Magister Johann von Kunsperk

(Johannes Regiomontanus.)

Figure 6.1 First page of a calendar compiled by Regiomontanus in 1473. This is the page for the month of January [Janer]. The second column on the left lists the dominical letters: A, B, C, D, E, F, G, from which the days of the week can be identified. Then the days of the Roman calendar are listed in the following column. The wide central column lists saints' feast days. This is followed by a double column indicating the position of the sun. Finally, two double columns designate the moon's longitude.

Figure 6.2 A German woodcut calendar listing the most important days for the years 1478–1496. The left-hand column lists the Golden Numbers used to identify years. The squares adjacent to the Golden Number column contain the dominical letters relating the days of the week and the intervals between feast days.

✠ 7
A Glimpse of Fifteenth-Century Life and Mathematics

Trade and Industry

The Treviso Arithmetic is a commercial reckoning book and, as such, reflects on the mercantile activities and concerns of its time. Its problem situations with their varied commercial scenarios convey a sense of economic dynamism that was very much an aspect of Venetian life. Commodities diverse and exotic passed through the hands of merchants, and fortunes were made or lost. Doge Tomaso Mocenigo, in a dramatic death-bed speech (1423), voices pride in the accomplishments of his Republic and, in particular, describes the financial tempo of Venetian trade:

> The Florentines bring to Venice yearly 16,00 bales of the finest cloth which is sold in Naples, Sicily and the East. They export wool, silk, gold, silver, raisins and sugar to the value of 392,000 ducats in Lombardy.

> Milan spends annually, in Venice, 90,000 ducats; Monza, 56,000; Commo, Tortona, Novara, Cremona, 104,000 ducats each; Bergamo, 78,000; Piacenza, 52,000; Allesandrea della Paglia, 56,000; and in their turn they import into Venice cloth to the value of 900,000 ducats, so that there is a total turn over of 2,800,000 ducats. Venetian exports to the whole world represents annually ten million ducats; her imports amount to another ten million. On these twenty millions she made a profit of four million, or interest at the rate of twenty per cent.[1]

Venetian commerce and the trade that passed through Treviso was indeed extensive. Although the Treviso book does not mention specific destinations for the wares it considers, many Italian commerical arithmetics did. Tartaglia, in his book of 1556, included monetary exchange rates between Venice Rome, Naples, Lyons, Antwerp, London, Paris, Milan, Pisa, Perugia, Bologna, Genoa, Florence, Valencia, and Palermo.[2] Rudolff is more specific in his work and actually mentions, by name, the trade items exported northward to the Germanic cities; for example, cloves and saffron to Nuremberg, soap to Augsburg, and cloves to Vienna.

The scope of the trade items mentioned in the *Treviso* is broad, and includes saffron, pepper, cinnamon, ginger, sugar, wheat, silver, cotton, crimson cloth, French wool, balsam, and wax. Such a listing reflects the modus operandi of the average Venetian merchant. At this time, merchants did not limit their trade to single commodities, but, rather, bought and sold a variety of products, thus ensuring their investments both a fluidity and mobility. The wide range of trade items held by a particular merchant is illustrated by the estate records of one Venetian, Pietro Soranzo, whose warehouses upon his death in 1350 were found to contain pepper, nutmegs, cloves, tin, lead, iron, Russian gold, raw silk, Russian furs, Syrian and Cypriot sugar, wax, honey, and pearls.[3] A predominance of textile and textile-related materials figure in Venetian trade—wool, cotton, crimson cloth, and yarn cloth—and attest to the area's emergence as a textile-producing region. While Venetians first served as middlemen for Europe's textile industry, importing cotton from abroad, transporting wool, and selling finished cloth, they eventually realized that the processing and manufacture of textiles, combined with export, would be even more lucrative. Thus, the Venetians began to dye cloth using the very dyestuffs they imported. Brazilwood, called *verzino,* which came from India, produced a very vivid red for which the Venetian dyers became famous. Silk, originally an Eastern import, also became a Venetian manufacture.[4] Lane, in his detailed study of Venice, points out how a silk weaver would subcontract work at 16 to 20 solidi a yard for cloth for which he would eventually be paid 30–32 solidi per yard and, thus, realize a handsome profit.[5] By the sixteenth century, Venice had emerged as a center for the manufacture and processing of woolens. The Venetian wool industry was complex and ruled over by a guild, *Camera del Purgo*. Wool, upon reaching Venice, was graded, washed with alum, combed, and spun into thread. This spinning was usually accomplished by women, *filatrice,* in villages around Venice. Spinning comprised a cottage industry. Finished thread was then warped (i.e., measured off to correct thread length to form the warp or web of a piece of cloth), by specialists, *orditori,* sized, and dried. It was then delivered to a weaver, *tessitori,* who made it into cloth. The cloth was then fulled, *follare,* washed, and softened at Treviso, given to teaselers who raised a nap on it, then sheared into a smooth surface, and finally dyed to a desired color. Thus, the process supported many artisans and the woolen industry became an important component of the economy.[6]

Similarly, other industries blossomed in Venice after their products were first introduced as import commodities. Fine Moroccan and Cordovan leathers were imported into Europe through Venice, but eventually the Venetians established their own tanneries and processed hides from Alexandria. Almost all Italian arithmetics of this period mention wax as an item of trade. Candles, of course, were the main source of illumination used by the upper classes, thus stimulating a demand for clean-burning beeswax. Venetian merchants imported this wax and developed their own factories in which the wax was whitened and formed into candles. Pearl buttons manufactured in Venice became a fashion rage of Europe. Other important Venetian industries that originated in such a manner were printing, glass blowing, and the minting of coins.[7]

Venice, quite early, established a dominance over the Eastern spice trade and held a monopoly on it until the end of the fifteenth century when competition from Dutch and Portuguese merchant-adventurers opened the sea routes to India and China. Trading in spices was extremely lucrative for two reasons: spices compacted into bales or baskets that were easily transportable, and there was a waiting market for them. For the Europeans of the fifteenth century and earlier, spices were almost magical substances employed to brighten up one's daily life. They served as seasonings, confections, medicines, and aromatics.[8] An example of the tastes of this period is given by the cargo manifest of a Venetian galley unloaded in London in the year 1420; it included ginger, saffron, mace, cinnamon, and rhubarb.[9] The principal spices popular in the Middle Ages were pepper, cloves, saffron, nutmeg, mace, cassia, and ginger. Cloves and nutmegs were imported from the Moluccas, pepper from southern India, Java, and Borneo, and ginger from Malabar. Of these spices, perhaps pepper was in greatest demand. The value of pepper has a long history in Europe. Persius (A.D. 34–62), the Roman satirist, noted:

> The greedy merchants, led by lucre run
> To the parch'd Indies and the rising sun;
> From thence hot Pepper and rich Drugs they bear,
> Bart'ring for Spices their Italian ware. . . .[10]

When, in the fifth century, Alaric, King of the Goths, besieged Rome, he demanded 3000 lbs. of pepper as partial ransom for the city. Pepper was a primary ingredient in the preservation of, and eventual seasoning of, meat. At this time in European history, it was the practice to butcher beef herds in the fall of the year while keeping a small amount of breeding stock to replenish herds for the following year. The meat was dried, salted, smoked, or frozen throughout the winter and consumed into the following summer. Pepper played an important part in the preservation process. To say the least, meat preserved in such a manner was frequently bland, if not tainted, and required heavy seasonings to increase its palatability. Pepper, cloves, and ginger were used mainly for this purpose. The demand for pepper was great, and a cargo of the spice could easily double the expended investment. In some years, Venetian pepper imports into western Europe amounted to more than a million pounds.[11]

No single substance is mentioned more frequently in early arithmetics than saffron, the red stigma of the *Crocus Sativus,* an ephemeral flower that blooms for less than a week before it dies. When the stigma is dried and ground, it produces a bright yellow powder, saffron. In ancient times, saffron was considered a sacred substance and was associated with religion and royalty.[12] Solomon's Song of Songs mentions the spice, as does the Koran. Hindu and Buddhist holy men colored their robes with the powder, and even today, its golden hue marks royalty in several Eastern societies. In the West, Nero on his coronation day (54 A.D.) had saffron strewn in his path upon entering Rome. An object of trade in China, Persia, and North Africa, saffron was introduced into Europe by returning crusaders, and became especially popular in the fifteenth and sixteenth centuries as a dye, a medicinal herb, a poison, and a cosmetic. Taken as a medicinal remedy, the spice was believed to cure melancholia and toothaches and to induce labor in pregnant women. Elegant ladies of Venice and other centers of European fashion adopted it as an eyeshadow. A lady's toilet in the fifteenth century required ingenuity and courage.[13] Saffron provided a striking and unusual cosmetic. The spice also served to a minor extent as a coloring agent in cooking, as we are told in Shakespeare's (1564–1616) *Winter's Tale,* "I must have saffron to color the warden pies."[14]

In an era when odors abounded, the use of aromatics became a social nicety. Nutmegs were used to fumigate the streets of Rome before the coronation of Henry VI. Baths of noble ladies would be scented with cinnamon, as would the bridles of mules used in a wedding procession. Geoffrey Chaucer (1340–1400) in his *Canterbury Tales* relates how:

> They danced and drank and, left to their devices,
> They went from room to room to scatter spices about the house.[15]

Spices were also used as medicines. Warren Dawson, in his *Collection of Medical Recipes of the Fifteenth Century,* notes that 26 spices found their way into medical prescriptions.[16] For many years, spices were thought to provide protection against contracting contagious diseases. Giovanni Boccaccio (1353) in writing on the Black Death commented:

> In this sore affliction and misery of our city [Florence] . . . many others went about, carrying in their hands, some

flowers, some oderiferous herbs and others some divers kinds of spiceries, which they set often to their noses, accounting it an excellent thing to fortify the brain with such odors, more by token that the air seemed all heavy and attained with the stench of the dead bodies and that of the sick and of the remedies used.[17]

Society

What can the passages of this reckoning book tell us about the daily life of the people at this time? One immediate question that might arise and that can easily be answered by considering the contents of problems is: For whom were the items of trade intended? Although Venice's position as a trade entrepôt benefited all its citizens by providing jobs and ensuring a flow of wealth within the society, most of the items passing through her custom houses were destined for a privileged few; they were luxuries by the standards of the fifteenth century. In the wake of the Black Death, Europeans exhibited an enhanced demand for luxuries. Artisans and merchants directed their efforts towards satisfying this desire. Long, hazardous voyages to foreign ports in search of merchandise were warranted by ample financial returns, which could be realized by catering in large part to the needs of the wealthy.

Due to fluctuating monetary standards, the comparative value of Venetian trade goods would be difficult to gauge by present-day criteria, but could be compared to values set for the fifteenth century. For such a comparison, let us consider the wages or income levels for various strata of society: an unskilled laborer probably made from 15 to 20 ducats a year; a journeyman builder, 50 ducats a year; a skilled foreman shipwright, 100 ducats a year, or $8\frac{1}{3}$ ducats a month; a nobleman of good economic standing would have had an income of around 1,000 ducats a year, and a very wealthy merchant or entrepreneur, 10,000 ducats a year.[18] Venetian tax records from the year 1581 reveal that 59 heads of households had incomes over 2,000 ducats.[19]

Thus, pepper selling at a little over 8 ducats per hundredweight wholesale would probably only be within the purchasing power of the wealthy. Beef, selling at 1 grosso for 3 lbs. in the fourteenth century, was most likely out of the price range of the lower classes, and probably was limited by cost even for middle-class people.[20] Saffron, selling for 7 lire de piccoli a pound, or a little over a ducat a pound, was a luxury solely for the rich. Similarly, the wholesale prices of other commodities such as wax, at 5 ducats per hundredweight; sugar, at 6 ducats; and ginger, at 8 ducats per hundredweight, limited these items to the larders of the noble and wealthy.[21] The most expensive organic substance considered in the *Treviso*'s problems is balsam, the aromatic resin from a tree, at 150 ducats per peso, a fraction of an ounce. Used as a medicinal substance and the base for some cosmetics, it demanded a rather high price at this time.

Spices and their use were associated with the rich. In times of social-class unrest, this fact was often brought to the fore:

> By what right are they whom we
> call lords greater folk than we?

> They are clothed in velvet, and warm in their furs, while we go covered with rags.
> They have wine and spices and fair bread, and we eat oatcake and straw and have water to drink.[22]

Even the cloth people purchased and the clothes they wore were a symbol of their social status. High-grade wool from Morocco, Spain, and England was reserved for use by the upper classes.[23] Crimson cloth, at 5 ducats a yard, is certainly not intended for the clothing of common people. Although locally dyed and processed, the price of this commodity placed it out of the hands of all but the very wealthy. Crimson, in particular, was a color associated with authority and high social standing. The Doge often sported a crimson mantle, and the official robes of the Chief of the Council of Ten and the Grand Chancellor were red, in contrast to the long black robes favored by the Venetian merchant class. Elegance of dress among the nobility achieved a high standard in the late Middle Ages and early Renaissance. Venice boasted a fraternity of young noblemen, dandies, called "The Hose":

> The young bloods, members of the club, wore fancy doublets of silk velvet, embroidered in gold and fitting tight to the body, with a belt at the waist. The sleeves were slashed, but tied together at points with ribbons, leaving puffs of white shirt to come through. The hose were tight fitting and striped lengthwise in colours; the shoes were pierced at the toes; on the shoulders a cloak of cloth-of-gold, damask, or crimson velvet, with a hood on the lining of which was embroidered the device of the club.[24]

Court life reached a peak of opulence in fifteenth-century Italy. Despite the elegance of individual attire, the acquisition of works of art, and the amassing of great private libraries, standards of personal etiquette, even among the high-born, were still being refined. Giovanni della Casa (1503–1556), Secretary of State to Pope Paul IV, in counseling his nephew on social graces, advises:

> Your conduct should not be governed by your own fancy, but in consideration of the feelings of those whose company you keep. . . . For this reason it is a repulsive habit to touch certain parts of the body in public, as some people do.
>
> When you have blown your nose, you should not open your handkerchief and inspect it, as if pearls or rubies had dropped out of your skull.
>
> It is not polite to scratch yourself when you are seated at table. You should also take care, as far as you can, not to spit at mealtimes, but if you must spit, then do so in a decent manner.
>
> It is bad manners to clean your teeth with your napkin, and still worse to do it with your finger. . . .
>
> It is wrong to rinse your mouth and spit out wine in public, and it is not a polite habit . . . to carry your toothpick either in your mouth, like a bird making its nest, or behind your ear. . . .
>
> No one must take off his clothes, especially his lower garments, in public, that is, in the presence of decent people. . . .
>
> Anyone who makes a nasty noise with his lips as a sign of astonishment or disapproval is obviously imitating something indecent,

and imitations are not too far from the truth.[25]

Prices for several textile goods are given, attesting to both the variety and grade of commodities: French wool at 120 ducats/1000 lbs.; wool at 19 ducats/hundredweight; silk at 42 ducats/hundredweight (obviously raw silk); yarn (we are not told what type) at 18 ducats/hundredweight; and cotton at 36 ducats/hundredweight. Linen, an important item in the Venetian economy, receives no attention in the *Treviso Arithmetic*'s problem situations; however, other arithmetics of this time tell us that linen was baled in Venice and exported overland on mules.[26] Spanish linen was priced from 94 to 120 ducats/hundredweight, Italian linen cost as much as 355 ducats/hundredweight, Saloniki linen, 380 ducats, and French linen was rather cheap at 140 ducats/hundredweight.

One problem given in the *Treviso Arithmetic* that attests to the plight of the common folk concerns the price of bread or, rather, the price of a bushelful, *staro,* of wheat.[27] A price of 8 lire for a bushel of wheat appears to be famine prices. Bread problems were standard inclusions in early arithmetics. Loaves of bread were of two types: "assized bread", which was always sold at a fixed price but varied in weight according to the price of wheat, and "prized bread", which was always the same weight although fluctuating in cost. The *Treviso* problem concerns "assized bread", whose standards were probably determined by law. Legal regulations concerning the weight of loaves of bread, the common man's food, are found in the Frankfort Capitulary of 794 and can be traced back further in history to Roman law. Famine often attended crop failures. In times of economic hardship, peasants frequently had to surrender their wheat in payment for rents and taxes and were forced to make bread of millet. The famine of 1527–1529 saw the city of Venice denude the countryside of wheat. In turn, the city was inundated with starving peasants, as described by an observer:

> Give alms to 200 and as many again appear, you cannot walk down a street or stop in a square or church without multitudes surrounding you to beg for charity: you see hunger written in their faces, their eyes like gemless rings, the wretchedness of their bodies with skins shaped only by bones. . . . Certainly all the citizens are doing their duty with charity—but it cannot suffice, for a great part of the country has come hither, so that, with death and the departure of people, many villages in the direction of the Alps have become apparently uninhabited. . . .[28]

It was not until corn (maize) was introduced from the New World and its cultivation became popular in the seventeenth century that the ever-present threat of famine was eased.

Money and Finance

From the contents of various problems given throughout the arithmetic book, it appears that a reckoning master working in the commercial climate of this time would be involved with diverse monetary calculations. The use of money as a means of inter-regional exchange was in its adolescence and experiencing growing pains.

A brief review of the development of money economies in Europe will help one to understand the situation customs clerks and commercial reckoners found themselves in during the fifteenth century.[29] With the fall of the Roman Empire, the coinage of Rome, no longer backed by the authority of the Senate or the Caesars, became merely a commodity rather than a unit of exchange associated with some distinct value. An item could be obtained for so many coins or as easily for another item of relative value. Barter, to a large extent, replaced a monetary economy. Gradually, from the fifth to the seventh century, gold coins from the Byzantine Empire became popular in the Mediterranean region as a medium of exchange. The Greeks called these coins *nomisma*. By the end of the seventh century, Byzantine influence in this area was replaced by that of Islam, and the Muslim *dinar* became the predominant currency of trade. With the rise of Italian capitalism, the golden *Fiorino* of Florence appeared in 1252 and, backed by the commercial strength of its city-state, became the standard for international transactions. As Florence's economic power was superseded by that of Venice, the Venetian *ducato,* ducat, or *zecchino* first appeared in 1284 and challenged the financial position of the *Fiorino*. The ducat was of such quality, both in the preciseness of its weight (3.59 gm) and the assayed value of its gold, that it became the monetary standard for trade throughout the known world and retained this status for many centuries.[30] The name "ducat" literally means "coin of the duchy or dukedom." Venice's doge was, in reality, a duke, and the coin's name is said to have evolved from the initial motto it carried, *"Sit tibi Christe, datus auam tu regis, iste ducatus"* (Let this duchy, which Thou rulest, be dedicated to Thee, O Christ).

Only under the rule of a powerful central authority could a value be established and maintained for money. In the Middle Ages, Carolingian reform set standards for monetary value based on silver. A pound *(libra imperiali)* of pure silver was to be struck into 240 pennies; the existing unit of soldi (shillings) then became worth 20 pennies. The penny of Charlemagne was set at 1.7 gm of silver.[31] Thus, a standard to be followed by European monies was set: 1 pound silver = 20 shillings = 240 pennies.

The Venetians followed this sytem, but on the eve of the fourth crusade Doge Enrico Dandolo introduced the denari grossi, or "large penny." This grossi was 96.5% pure silver and weighed 2.18 gm. The existing penny now became the "small penny," piccolo, and two different monetary units formed a pound: lira di piccoli and lira di grossi. While the two different pounds existed side by side, they differed in value by the ratio 1:26.9 (see Table 7.1).

The large units of money became the standards for international trade and the basis of monetary value. Fractional denominations existed for the convenience and use of the common people. Throughout the Middle Ages, hard currency was "aristocratic money." Only

Table 7.1 Monetary System of Fifteenth-Century Venice

Lira di piccoli =	20 soldi =	240 piccoli =	19.33 gm silver
soldi di piccoli =		12 piccoli	
piccoli (pizoli)			
Lira di grossi =	20 solid =	240 grossi =	504.72 gm silver
soldi di grossi =		12 grossi	
grossi =	32 pizoli		
Ducat =	24 grossi		
10 Ducat =	Lira di grossi		

the wealthy moved and accumulated sums of money, but gradually by the fifteenth and sixteenth centuries, due to the growth of local industry and commerce, the use of and access to money became "democratized." Increased monetary demands at this time placed strains on the supplies of denominate coinage—there was a shortage of small coins. Periodic debasement of the fractional parts of the monetary units resulted in frequent re-evaluations of monetary equivalents. Many mints were private enterprises; therefore, standards of minting varied. Treviso, at one time, had its own mint. But, despite the unreliability of the quality of minting in general, the mint at Venice was recognized throughout Europe for the high standards of its work and was commissioned to coin foreign money and forge silver ingots for Hanseatic towns and the merchants of England.

Furthermore, actual physical debasement of silver and gold coins by coining, chipping, or sweating was not an uncommon practice. Coins were so marred and defaced that money changers actually weighed out coins by the pound; thus, a *lira di piccoli* would actually be a pound of silver pennies and possibly vary in number from one exchange to another. In order to preserve the integrity of its money, the Venetian Republic enacted rather stringent laws intended to discourage would-be offenders: in 1321 provisions were made to inspect money changers' benches to insure against debasing; in 1357–1359, the penalty for this offense was the cutting off of the right hand; those caught coining were physically disfigured—women had their noses cut off and men were blinded.[32] Clandestine mints circulated bogus coins. Thus, at this period of history, money and its use was still a questionable matter in many respects.

Figure 7.1 A 16TH-century woodcut print by Hans Weiditz of Augsburg depicts a money changer at work in his bank. Usually, such money changers were also merchants.

In an era with such a diversity of currencies, the mathematics of currency exchange was an important aspect of commercial training. The *Treviso Arithmetic* contains only two problems that specifically treat monetary exchange—the changing of florins to ducats and to Rhenish florin. Perhaps the author only intended to introduce the mechanics of the process and thought it unnecessary to develop an extended list of exchange problems; however, Tartaglia felt this aspect of commercial arithmetic so important that he devoted many pages of his book to exchange. For example, he considered the exchange rates between Venice and fourteen other cities and between Florence and twenty-one cities.[33] A specific procedure, the chain rule, *regula del chataina,* for computing exchange through various cities existed in the computing schemes of the time. It was introduced into Europe from Arab sources via Leonardo of Pisa. Consider the following problem from Riese: 7 pounds at Padua make 5 pounds at Venice, 10 pounds at Venice make 6 pounds at Nuremberg, and 100 pounds at Nuremberg make 73 pounds at Cologne; how many pounds at Cologne do 1000 pounds at Padua make?

The entries are arranged according to the following scheme:

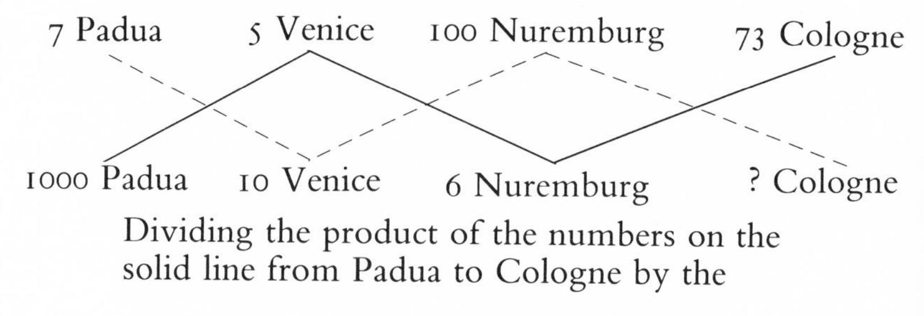

Dividing the product of the numbers on the solid line from Padua to Cologne by the product of the numbers on the broken line from Padua to Cologne,

$$\frac{1000 \times 5 \times 6 \times 73}{7 \times 10 \times 100} = 312\frac{6}{7}$$

thus, 1000 pounds Padua = $312\frac{6}{7}$ pounds Cologne. No mention is made of this technique in the *Treviso* work.

While the reckoning of exchange rates, tariffs, and allowances was often confusing and difficult, this situation did not hinder the Italian merchants' quest for increased revenues. In fact, if one word could be chosen to describe the motivation of merchants at this time, it would be *avanzo,* profit, as they, themselves, gladly admitted. A merchant of this period, Francesco di Marco Datim, inscribed the headings on each page of his ledgers "in the name of God and Profit."[34] The spirit of capitalism reigned supreme. It influenced all transactions, as noted by the advice of Andrea Barbarigo to his agent, or *fattore,* in the year 1440:

> I think they [the clothes shipped] will bring 30 ducats if sold for payment in six months and 28 ducats if sold for cash. If you cannot sell them for cash at 2 ducats less than the term price, I even prefer that you sell them for cash at 3 ducats less, because one knows how to gain more than 12 percent a year on one's money.[35]

Profit also indirectly served as a stimulant for the various business, financial, and mathematical innovations instituted by the Italian merchants. Whether it be the employ of double-entry bookkeeping procedures, the use of the "letter of exchange,"[36] the adoption of maritime insurance, or the application of the Rule of Three in the solution of a business problem, all such schemes were intended to save money, protect investment, and increase profits.

Easy profits could be obtained through the lending of money at high interest rates. Interest rates varied with the times and circumstances; for example, a high-risk fourteenth-century English king borrowed money from international merchants at an interest rate of 260% per annum.[37] On the average, they were usually high as judged by present-day standards. The Florentines had a saying that summed up the situation concerning the lending of money: "25 percent interest amounted to nothing, 50 percent would do to pass the time while 100 percent would prove interesting."[38] But despite the inducements of money lending in search of profit, the merchants of Italy had to be careful not to fall afoul of the Church's usury laws.[39]

As has been previously discussed, the financing of business ventures was often undertaken in partnership with family members or fellow merchants. Such mergers of funds allowed capital to earn interest without resorting to lending money on the open market and indulging in recognized usurious practices. Other devices were also utilized to circumvent the Church's usury laws, the simplest perhaps being the adoption of a variety of euphemisms for the word "interest": *prode* (yield), *guadaguo,* (gain), and *merito* (reward).[40] Christian businessmen, while allowed to invest capital in trade ventures, could not freely lend funds on the open market, but still there were times when money had to be obtained quickly and with minimum social and political constraints. To satisfy this need, Jews, not being bound by the Church law, were allowed to be money lenders. This concession was mutually agreeable to both the Christian and Jewish communities. Jews were limited by discriminatory laws to certain professions and money lending provided an opportunity for lucrative gains, as well as direct participation in commercial life. Treviso, itself, had a large community of Jewish money lenders who did business with Venetian merchants. The Jewish community of Venice was limited to one section of the city, *Ghetto Nuovo* (*ghetto* means casting, *Ghetto Nuovo* literally means "new foundry"), and its money lenders became famous. It is with these money lenders and their benches, *banco,* or tables, *travolo,* of business, that the origins of modern banking began.[41] When a money lender was found to be cheating he was publicly disgraced and put out of business by having his bench broken, *banco ruptus;* today we would say he went bankrupt. The merchants and money lenders of Venice are immortalized in Shakespeare's play *The Merchant of Venice* (c.1595).

Weights and Measures

In the West, the fifteenth and sixteenth centuries mark the beginnings of large-scale international trade. The Venetians were well-established in this field with a string of colonies around the Baltic and along the coast of North Africa. Overseas distribution centers for Italian goods were called *fondaci,* under the direction of a trading agent, a *fattore.* The chain of European trading colonies extended as far eastward as Goa, on the coast of India, and Malacca on the west coast of what is now Malaysia. On continental Europe, trade fairs served as the focal point for large-scale commodity exchange. During the first half of the fifteenth century, in the region around Paris alone, 32 trade fairs were held.[42] The towns in which the fairs were held, themselves, became centers of trade and influenced trading practices. Many Italian mercantile houses maintained establishments at Lyon and the wholesale traders became resident merchants with eventual branches in all the great cities of Europe. As the scope of trade expanded geographically, the need for systems of standardized weights and measures became more obvious. Each region had its own system of measurement, as testified to by the report of a seventeenth-century European traveler who noted the word for the weight measure pound specified 391 different units, and the word for foot, 282 unlike units by his count.[43]

Robert Lopez in his examination of a Pisan merchant's manual of the late thirteenth century, *Memoria de tucte le mercante,* noted apparent discrepancies in the exchange of weights and measures between the various trade destinations listed. For example, Lopez discovered that 100 *salme* of wine from Scalea, in the southern Italian region of Calabria, became 95 *mezzaruole* when exported to Tunis, yet 100 *salme* of another Calabria wine corresponded to 82 *mezzaruole;* a *centenaro* of hazelnuts from Naples was measured at between 150 and 155 *cantari* upon reaching Tunis and 145 *cantari* in nearby Bougie, whereas a *centenaro* of walnuts from Naples became 140 *cantari* in both Tunis and Bougie.[44] In such a melange of exchanges, a merchant welcomed information on foreign measures that bore similarities to the standards he used and made careful memos of the fact: "In Morea people weigh by the Lucchese pound, which is one fifth an ounce larger than the Pisan one."[45]

A system of weighing jewels and precious metals was devised at Troyes, a center of trade for precious items in the Middle Ages. Since Venice controlled much of European trade, in a similar manner many Venetian weights and measures served as general standards for European and near-eastern trade. The *Treviso* presents a variety of such measures.

miara, thousand weight = 1000 pounds
centonaro, hundredweight = 100 pounds
lira, pound = 12 ounces
onza, ounce = 300 pexi

While this grouping of weights seems straightforward enough, it should be noted that there also existed *migliaio grosso,* a great thousandweight, used in Venice and Constantinople; *lira grosso,* a heavy pound; and *lira sottile,* a light pound, and that a hundredweight of the time was a rather relative measure, usually being more than 100 pounds (for example, in England it was 112 lbs.).[46]

The *Treviso* mentions the *sazi,* a weight for spices equivalent to $\frac{1}{6}$ oz.; however, it does not acknowledge the standard quantity of pepper, the *sporta,* about 600–700 lbs. Precious metals are weighed by the *marche, onze, quarteri* and *carato,* where *marche* = 8 *onze.*

onze = 4 *quarteri*
quarteri = 36 *carato*[47]

The only dry measure given in the text is the *staro,* bushel, which became the base for a multitude of grain measures: the *carica* was 6–7 Venetian bushels in Beirut; *carro,* 22 bushels in Apulin; *varra,* 6 Venetian bushels in Seville; *ribeba,* $3\frac{1}{2}$ bushels in Alexandria; *quarta,* $\frac{1}{4}$ bushel; and *quartarolo,* $\frac{1}{16}$ bushel.

Cloth is measured in *braza,* yards which comes from *braccio,* arm's length. The Venetian term for mile is *miglia,* which evolved from the Latin prefix for 1000, *mille;* a Roman mile, *mille passus,* was 1000 paces, as is the Venetian mile. The Venetians also used pace, *passa,* as a measure of distance, as given in the hound and hare problem of F55v.

The *Treviso*'s listing of measures, while not extensive, still convey two impressions concerning the state of weights and measures at this time: a mathematical base of ten was firmly established for quantitative comparisons, and the world of measurement was complex and often confusing in its scope and variety.[48]

A Mathematical Testament

The *Treviso Arithmetic* reveals much more than a few mathematical techniques for solving problems. Neither elegant nor necessarily outstanding by the standards of the fifteenth century, the book and its contents still have an enduring value. There are times in history when the coming together of several forces or events produces an effect far surpassing that which could have been produced by each alone. It is the events of such a period of history that the *Treviso Arithmetic* reflects: printing, the popularization of the Hindu-Arabic numerals, and the rise of mercantile capitalism. While the Treviso tome directly speaks of mathematics, it also tells a very human story of people, their society, and the changes they are undergoing. The true title of the book, if it had one,[49] remains unknown, as does the identity of its author, yet their influence is neither lessened nor muted by anonymity.

The author is a *maestro d'abbaco,* relatively speaking, a learned man of his time. He is familiar with the works of Boethius and other Latin writers and is an experienced reckoner, skilled in the computational practices of the day. But perhaps the trait that emerges most clearly from this book is that its author is innately a teacher concerned with the transmission of knowledge to his fellow man. A personal relationship is established between the author and his readers and it is maintained throughout the text. As we were told, the author was sought out and requested, by certain youths who aspired to commercial careers and for whom he had some affection, to write the book.[50] What an admirable situation! In the instruction that unfolds, a paternal tone is frequently heard. The author encourages: "Attend then diligently to the arrangement of the operations I set forth"; flatters: "Those who are of scholarly tastes should learn . . . " and admonishes his charges, "What availeth virtue to him who does not labor? Nothing."[51] An interesting comparison of the instructional style of the Treviso reckoning master with that of a university professor of the time can be made; Philip Melanchthon, in lecturing his mathematics class at Wittenberg (1517), said:

> Now that I have discussed for you the usefulness of the art of computation, of which there cannot be the slightest doubt, I believe that I should also make some brief remarks about the ease with which it can be done. I believe that students allow themselves to be frightened away from this art because of their preconceived notion that it is too difficult. As far as the elements of computation are concerned, which are already generally taught in schools and used in daily life, those who think that they are too difficult are greatly in error. This knowledge springs directly from the human mind and appears with full clarity. Therefore its elements cannot be obscure and difficult; on the contrary, they are so clear and evident that even children can grasp them, because everything proceeds so naturally from one point to the next. The rules for multiplication and division, to be sure, require more diligence for their

> mastery, but their meaning will still be understood very quickly by those who give to them their full attention. These skills, of course, like all others, must be sharpened by practice and experience.[52]

The fact that the *maestro d'abbaco* author has been sought out by students attests to both his respected status in the community and the attractiveness of the reckoning master profession. Reckoning masters of this time are the forerunners of the applied mathematicians of a later date; their calculations spanned a multitude of disciplines from commercial reckoning to land survey, calendrical computation, and barrel and cask gauging. Here, in the fifteenth century, the practicing of mathematics is recognized and esteemed in society as a profession unto itself. Arithmetic is no longer considered a mere novelty or conveyor of Neoplatonistic mysticism as it was in the Middle Ages. Neither are mathematical "secrets" limited, either by design or circumstance, to an intellectual fraternity of clergy, academicians, or privileged nobles. Nor are mathematical symbols and techniques considered dark and nefarious devices associated with the powers of evil, as they once were:

> The good Christian should beware of mathematicians and all those who make empty prophesies. The danger already exists that mathematicians have made a covenant with the devil to darken the spirit and to confine man in the bonds of Hell.[53]

The demand for the skills of knowledgeable computers eliminated the exclusive status of mathematics. This condition, fortified by the appearance of printed texts, economically attainable by the bourgeois, ensured a popularization of arithmetical and mathematical knowledge. Printing facilitated the dissemination of hundreds of copies of the same text. Under the economic and intellectual impetus of this time, not only were mathematical techniques being more widely learned but they were, in many cases, new techniques based on the use of Hindu-Arabic numerals and their accompanying algorithms. From this period onward, computation involving numbers would be more easily executed and efficiently recorded. The visual stimulus of a mathematical process written out allowed for a re-examination and questioning of the process; patterns could be noted and mathematical structure discerned. Printing also forced a standardization of mathematical terms, symbols, and concepts. The way was now opened for even greater computational advances and the movement from a rhetorical algebra to a symbolic one. Thus, the information of the *Treviso,* in a very real sense, marks a turning point in mathematical thought and potential.

A modern reader of the *Treviso Arithmetic* may be disturbed by its rather terse style. The book was, after all, a *practica,* written to convey basic knowledge and techniques in very simple terms. In this sense, it is a manual expounding 'do this and you will get this result.' Little consideration is ever given to the 'why' of what is taking place. We, being educated in a system where theory and practice are taught in conjunction, find it difficult to appreciate so skeletal an approach to teaching arithmetic, but it must be remembered that by the needs and demands of the fifteenth century such an approach was not only adequate but desirable. Prolonged schooling was expensive. A youth wishing to learn commercial arithmetic acquired the basics from a master such as our author and then entered the profession in a minor role as an apprentice in which he continued to learn. As his skill and experience increased, hopefully his inquisitiveness concerning the techniques of his art would grow. Thus, perhaps after several years of work and association with other reckoning masters, a computer might truly begin to seek out the 'whys' of arithmetic, as our author seems to do by questioning the efficacy of proof by 'casting out nines.'

The limited scope and intensity of the instruction can also be explained by two other justifications. No doubt the size of the book was an economic factor. Printing was a new art and the composition of type and diagrams for a mathematics text posed particular problems for early printers. This fact is duly noted by Douglas McMurtrie in his book on the history of printing, where he erroneously attributes the appearance of the first printed Italian mathematics book to 1482.[54]

The second reason for the brevity of the book is simply that the author was exhibiting the discretion of a good teacher. His text was written for a specific audience with needs associated with the region around Treviso. There was no intention of its being comprehensive, or a standard text on the subject of computation. Nor was there any reason for it to be an encyclopedic compendium of mercantile knowledge and mathematical techniques, as later composed by Pacioli (1494) and Tartaglia (1556). Neither was it a merchant's handbook of commercial problems and their solutions, as many practicae or *abaci* of this period have been characterized.[55] Certainly, it was not a study on or about arithmetic; this was left for the Latin writers, the academics of this period.

The *Treviso Arithmetic* was a book from which one learned mathematical knowledge, the symbols and techniques of arithmetic, and the methods of commercial reckoning, and developed some appreciation for the applications of that mathematics. Its author's methodology throughout would seem to uphold this conclusion. Instead of demonstrating eight methods of multiplication or six techniques of division, he built upon his experience and chose those algorithms he deemed best for the perceived needs of his charges. A similar decision seems to have been made concerning problem selection—a few well-chosen, representative problems are presented to provide practice on the basic operations and to synthesize the knowledge and techniques already learned.

From reviewing the contents of the book, two criticisms emerged: the treatment of fractions was sparse and fragmentary and the consideration of such topics as weights and measures, exchange rates, tariffs, etc.—specific mercantile information—was extremely limited as compared to that given in many other similar works of the period. The omission of a discussion on fractions can be explained in several ways. First of all, it is obvious that the reader is expected to have had a previous knowledge of fractions, as they are introduced and employed freely without additional explanations as to their nature and use. Why the author does not choose to expound on an arithmetic of fractions can be explained in two ways. First, he may have felt that a discussion of this topic in print would have been too confusing for the students and left it as a matter of personal face-to-face instruction. Second, and far more likely, he just did not feel that a detailed discussion of fractions was important for mercantile practice. This last statement requires an explanation. Traditionally, merchants circumvented the difficulties of fractional computation by introducing a multitude of composite units and subunits into situations such as monetary exchange (a quartarolo $= \frac{1}{4}$ soldo) or the allocation of a commodity (a quartarolo $= \frac{1}{16}$ staro), situations where the frequent division of some particular unit was required. When fractions did exist in some computation, they were often ignored, as Pacioli noted:

> Many merchants disregard fractions in computing and give any money left over to the house, thus committing fraud in their hearts and injuring the person who deals with them. I, myself, have been in certain cities of Italy where this evil custom obtains. [*sic*][56]

Whether a merchant would ignore a fraction when its reckoning would be in his favor is open to question.

The omission of extensive lists of currency

exchange rates or other such commercial information in the Treviso book is easily explained. It was redundant. This information was available in other books and manuscripts in the merchant's possession. Merchants were perhaps the first occupational group in European history to acquire professional libraries consisting of references and handbooks specifically written for their trade. These *practica della mercatura,* which began appearing in Italy in the thirteenth century, included such items as compilations of measures and systems of exchange for the various regions and countries trading with the Italians, (e.g., *Questo e ellibro che tracta di Mercatantie et usanze de paesi,* Georgio Chiarino [1481], or *Tariffa perpetua,* Giovanni Mariani [1535]), or collections of general information on trade procedures and customs that were of interest and value to merchants (e.g., *Memoria de tucte le mercantie* [1279]).[57] With the existence and availability of such detailed and comprehensive references, there was little need for repeating their contents in a text devoted to arithmetic.

It is interesting to note that even in a constrained instructional situation, that is, the presentation of limited material to a limited audience, several problems at the end of the text—the couriers, hound and hare, lost purse, and carpenters building a house—bear little "practical" use. These problems are not relevant to the commercial world and their origins certainly predate the Treviso writing. They are intellectual exercises and, in a sense, a boast of the power of mathematics. It is almost as if the author is saying to his readers, "Now that you've learned the basics, I'm going to show you something really interesting—some problems you can amaze and puzzle your friends with."

The early Renaissance has often been depicted as a rather stagnant period in the history of European mathematical development. Paul Rose, a mathematical historian, describes it as "a large featureless plain broken up by such occasional peaks as Cardano, Copernicus and Galileo."[58] Still other authors tell us that no significant mathematics appeared in Europe between the time of the death of Fibonacci (1250) and the beginnings of the sixteenth century.[59] Of course, such judgments are relative and rest on the interpretation of the word "significant." Significant for what and for whom? If one equates significance with the appearance of original contributions in the form of theories that advance the corpus of mathematical knowledge, then the sentiments of Rose are correct. But there is more to mathematics than theory, for theory by itself, without a mode of expression or articulation, remains impotent. While not dramatic in the sense of new theories, the mathematical accomplishments taking place in the fourteenth and fifteenth centuries are profound in a subtler way. These centuries mark an historical period of transition for Western civilization, a movement toward new intellectual traditions. Novel ideas and exotic practices imported from the East into Europe were taking hold. The conception of man and his place in society was changing. Man was no longer a passive observer. He became a 'doer' and a changer of his environment. It was for this end that

Europeans, particularly the Italians, began to use the Hindu-Arabic numerals and their associated algorithmic computational procedures.

Some scholars have proposed that the popular conception of the Renaissance as a time of 'rebirth' for Western ideals and ideas is fallacious. These notions were well on the way to exerting an influence by the time of the Middle Ages.[60] A humanistic and scientific spirit was emerging in Europe as early as the eleventh century[61] The period in which so much attention was focused on antiquity and the classics of Greece and Rome, which has been referred to as the "Renaissance," is but a phase in the broadening of human experience that was already taking place. From the thirteenth century onward, mathematics was being used more and more in the practical applications of daily life. It was only when the mathematics derived from these practical applications had reached some sort of maturity that it became responsive to the works of Diophantus and the theories of Apollonius. It is exactly this sense of emerging maturity that the *Treviso*'s contents convey.

The techniques and examples of the *Treviso Arithmetic* provide evidence on just how the use of Hindu-Arabic numerals had evolved and grown in Italy by the end of the fifteenth century. By that time the numerals themselves had evolved into the common form that is known and accepted today.[62] Similarly, the algorithms for addition and subtraction were standardized and conform to present-day practices. The processes of multiplication and division, however, were still in a state of flux, with several alternative algorithms available for use. Included among this latter variety of algorithms, and prominent within them, were the two basic techniques used today for multiplication and division. Fractions and fractional computation were cumbersome.[63] The significance of place value, while appreciated, and the use of the new numerals, had not been sufficiently established for its extension into a comprehensive theory of decimal fractions; this innovation would have to wait for the appearance of Simon Stevin's *La Thiende* (1585). The mathematical concept of percent was appreciated, and the value of making commercial transactions based on the unit of 100 seems to have been firmly established.[64]

The Merchant as an Intellectual and Scientific Innovator

If the *Treviso Arithmetic* testifies to the emerging maturity of mathematics, it also provides insights into the perpetuation and transmission of mathematical knowledge and the place of the merchant class in this movement. The role of merchants in medieval and early Renaissance society as instigators of economic, political, and scientific innovation is seldom fully appreciated. Careful examination and contemplation of the contents of the records left by these men are just beginning to reveal their importance in the scheme of historical accomplishments.[65] It was trade that initially opened Europe to the fresh ideas and knowledge that liberated it from the feudal mentality imposed on society after the fall of the Roman Empire. Trade opened new intellectual as well as physical horizons.

The driving force that brought Marco Polo, and his father and uncle before him, to the court of Kublai Khan was a search for profit. The Polos were Venetian merchants seeking to advance their fortunes but, as a result of their adventures and travels, European society learned of foreign lands and strange peoples. It is no mere coincidence that Fibonacci, one of the principal conveyors of a knowledge of the Hindu-Arabic numeral system to Europe, was also a merchant. During their travels, merchants were keen observers of foreign techniques, customs, and procedures that affected their trade prospects. They were also avid record keepers and compiled detailed logs of personal impressions and empirical facts relevant to commerce. Frequently, the contents of such logs were formalized into manuscripts or books of reference, i.e., the *practicae* of previous discussions. In a sense, Fibonacci's *Liber abaci* was a work of this kind, although it reflects the author's interests as a mathematician as well as a merchant. The contents of *Liber abaci* on the general nature of the Hindu-Arabic numeral system and its algorithmic procedures (chapters 1–7) and mercantile problems (chapters 8–11) formed the basis of a whole descendent genre devoted, primarily, to the mathematical needs of merchants.[66] This genre, which extended from the thirteenth through the sixteenth centuries, was, in itself, innovative. In producing a popular, rather than a scholastic, literature it championed the use of vernacular languages and introduced a practice of using pictures to illustrate mathematics problems.

The *Treviso Arithmetic* was a part of this tradition. Its opening statements leave no doubt of this fact as the author tells us that it was prepared for study by those seeking mercantile pursuits (*merchadantia*).[67] Other authors of commercial books employed different phrases to convey the same intent: "prepared for merchants", Borgi; "for business purposes", Riese; "for merchants, bookkeepers and beginners", Van der Schuere.

While the information and computational techniques considered in Fibonacci's *Liber abaci* appealed enough to merchants to ensure its preservation and transmission in their practicae as *abaci,* its adoption by European society, even among the merchant class, was not rapid. At first, the new knowledge seemed to have a great impact, but its spread in the thirteenth and fourteenth centuries was hesitant. Even in the fifteenth century, Pacioli lamented those merchants who still used "old and vulgar methods" in their arithmetic.[68] However, it was at the end of this same century, in the period of the *Treviso*'s publication, that the momentum for using the new arithmetic increased greatly.[69] For the merchant class, the late Middle Ages—the initial period in which they were exposed to the Hindu-Arabic numerals—was a time of extreme commercial opportunity and easy profit. As such, there may have seemed no need for the introduction of radical changes in the notations of record keeping and the techniques of computation. In some cases, vested interests further resisted reforms.

Gradually, a series of events changed the mercantile climate and forced a rethinking of business procedures. The collapse of the Mongol Empire and the rise of the Ottomans curtailed much Eastern trade. The Black Death and the ravages of the Hundred Years War reduced European commerce so that, by the fifteenth century, Europe was experiencing economic depression.[70] In this time of shrinking markets, smaller profits, and keener competition, merchants re-evaluated their situation and sought out more rational methods of operation. Thus merchants were once again attracted to the efficiency offered by the use of the new arithmetic.

In analyzing the temporal thrust of European arithmetic reform, it is interesting to trace and associate the appearance of commercial arithmetics with the rise of mercantilism and industrialization. For example, Italy's commercial progress is reflected by the publication of the Treviso book in 1478, Borgi's work (1484), followed by that of Calandri (1491), Pacioli (1494), and Tartaglia (1556). Hanseatic League trade opened the commercial possibilities of Germany and writers such as Weidman (1489), Riese (1518), and Rudolff (1530) responded to the increased demand for mathematical knowledge.[71] In the sixteenth century, the spirit of commercial adventure entered France, and Savonne (1553) expounded on the techniques of merchant arithmetic. The Netherlands awoke to her maritime potential and practical arithmetic appeared in the works of Van der Schuere

(1600) and Raets (1580). England's merchants, reacting to the continental tempo of trade, sought to establish their own markets and the writings of Recorde (1542) and Baker (1568) became popular.

Not only did mercantile activity foster a knowledge, use, and appreciation of mathematics, but it also encouraged an interest in exactitude and a concern for the accuracy of measurements of time, distance, and capacity. The merchants' involvement with the collection of data and record keeping—

> It is a most useful thing to know how to keep records properly; and this is among the principle lessons a merchant can learn.[72]

—and their quest for answers to problems from numerical data, elevated mere computation to the status of an empirical science. Thus these traditions were well-established by nonscientists well before the time of Galileo, Copernicus, and Descartes and their mechanistic world view. Further, the quest for new markets and greater profits also gave rise to studies in geography and astronomy and led to advances in cartography, navigation, and marine architecture.[73]

Some Final Thoughts

This brief consideration has exposed a link in the chain of historical transmission of mathematical knowledge and aids in the attainment of a better understanding of how mathematics and society are interrelated. While the concepts and techniques provided by the Hindu-Arabic numeral system were known in Europe for five hundred years prior to the appearance of the *Treviso Arithmetic,* their full development and exploitation lay dormant until a favorable social climate existed in which they could be appreciated.[74] This climate existed in 14–15th century Italy, particularly in Venice, where the spirit of boldness that was so much a part of early Venetian survival now turned to financial adventurism and the building of a commercial empire. Traders became international merchants, and the growth and reinvestment of wealth saw the rise of commercial capitalism and its associated institutions—a system of international monetary exchange based on the ducat, corporations and stock companies, and a system of deposit banking. As geographical and economic constraints were challenged and overcome, new intellectual and political horizons were also explored. Secularism and political independence from Rome provided a climate of freedom where unfamiliar knowledge of subjects such as arithmetic and mathematics was consciously pursued. In this period, a modern realization was dawning as to the usefulness and facility of arithmetic; these benefits gave mathematics a new value. While this vitalized appreciation first centered on arithmetic, for similar reasons it would soon spread to other areas of mathematical endeavor: geometry, trigonometry and algebra, and other sciences.

In an almost prophetic mode, the author of the *Treviso,* in his elegant manner, writes a final statement of encouragement to his readers.

> By as much study as you have given the work, by so much has it appealed to your ardent desires. I do not doubt it will bring back to you much fruit. Not that by such sacrifice, however, can we lead everyone to be learned or expert in this practice (since to them such teaching may not be necessary), but only to such as you who are desirous of education. And, therefore, according to your wish, if not wholly at least in part, and corresponding to my efforts by you so graciously recieved, I promise you the same gratifying usefulness.[75]

Indeed, today we are enjoying that usefulness explained and described by the fifteenth-century *Treviso Arithmetic*.

David Eugene Smith: A Tribute

The October, 1933 issue of *Science* notes that David Eugene Smith, Professor Emeritus of Mathematics, Columbia University, was personally awarded by His Imperial Majesty, the Shah of Persia, his government's Order of Elim. Smith was given this decoration in appreciation of his translation and publication of Omar Khayyam's *Rubaiyat* and in recognition of his interest in the history of Persian mathematics. The *Science* notice goes on to describe Smith's 1932 odyssey of four months and 10,000 kilometers across the Middle East in search of old Persian, Arabic, and Hebrew mathematical manuscripts. An unusual adventure for the times, an unusual award, and certainly an unusual man!

David Eugene Smith was born January 21, 1860 in Cortland, New York. Before David was twelve years old, he had learned Greek and Latin from his mother. He attended the newly-founded State Normal School in Cortland and advanced to Syracuse University where he majored in mathematics, art, and classical languages, obtaining a Bachelor of Philosophy degree in 1881 and a Master's in Philosophy in 1884. He went on to obtain his Ph.D in art history in 1887. Following his father's wishes, he became a lawyer, but only practiced in the profession briefly, abandoning law to become a teacher of mathematics at the Cortland Normal School. From there, in 1891, he went on to the State Normal College at Ypsilanti, Michigan, and eventually to Brockport, New York, as principal of its Normal School. In 1901, he was solicited for a faculty position at Teachers College, Columbia University. He accepted a professorship of mathematics and spent the remainder of his teaching career at Columbia until retiring in 1926.

It was at Teachers College that Smith embarked on the career that would truly distinguish him as a mathematical educator and historian of renown. Smith had already begun a writing career in Michigan when he published his first textbook *Plane and Solid Geometry* (1895), written in cooperation with W. W. Beeman of the University of Michigan. Beeman and Smith would collaborate on seven more secondary school texts; but at Columbia, Smith formed a writing partnership with George Wentworth that would produce the Wentworth-Smith Mathematical Series (1909–1922). This series eventually totaled 23 mathematics textbooks and dominated the American mathematics teaching scene for many years. Some of these books were translated into foreign languages and used around the world. Smith was not only a prolific writer of texts but also produced scores of professional articles, reviews, and pedagogical tracts. His principal scholarly works focused mainly on the history of mathematics, and included *Rara Arithmetica* (1908), *History of Mathematics,* 2 vols. (1923–1925), and *Number Stories of Long Ago* (1910). Several notable works were achieved in collaboration with partners: writing with Marcia Latham, he produced a translation of René Descartes's *La Géometrie* (1925); with Louis C. Karpinski, *Hindu-Arabic Numerals* (1911); with Yoshio Mikami, *A History of Japanese Mathematics* (1914); and with Jekuthiel Ginsburg, *A History of Mathematics in America before 1900* (1934). Smith's involvement with the learning and teaching of mathematics convinced him that in order to develop a good pedagogy for the subject one must know and understand its history. He devoted much of his life attempting to learn and understand that history and to convey it to others.

Smith was instrumental in the founding of the History of Science Society in 1924 and in 1927 served as its first president. He encouraged and supported the compilation and translation of historically-relevant materials. Many scholars such as Thomas L. Heath, Raymond C. Archibald, Eric T. Bell, Julian L. Coolidge, Vera Sanford, and Leonard E. Dickson benefited from his assistance. He also aided Otto Neugebauer in establishing the *Mathematical Reviews*, supported Herbert E. Slaught in developing the Carus Monograph Series, and played a primary role in the founding of *Scripta Mathematica* (1932). As an avid traveller, fluent in over eight languages, Smith journeyed widely, collecting rare mathematical books, manuscripts, and instruments. Eventually, he accumulated almost 20,000 books and items involving mathematics, mathematical activities, and mathematicians. His collection was particularly strong in materials from India and the Far East and medieval works from Europe and the Islamic world. He also assisted the scientific bibliophile George A. Plimpton in amassing an outstanding collection of mathematics books and manuscripts. Smith's collection was donated to Columbia University in 1931 and eventually became part of the Plimpton-Smith-Dale collection of the University's Butler Library.

During his lifetime, he served on many national and international committees and commissions devoted to improving the teaching of mathematics. Smith worked with Felix Klein on the International Commission on the Teaching of Mathematics (1908), a powerful stimulus for mathematical teaching reforms at the turn of the century. As a member of the Committee of Fifteen on the Geometry Syllabus (1909), he helped shape geometry teaching in the United States, and as a participant in the Commission on Mathematics of the College Entrance Examination Board he contributed to the shaping of our present-day secondary mathematics curriculum.

While his accomplishments were both profound and diverse, Smith remained primarily a teacher of mathematics. In this activity, he sought more than to merely teach the subject. He attempted to nurture in his students an almost spiritual reverence towards mathematics, one that transcended the classroom. As one anonymous biographer noted:

> The mere acquiring of ability to solve a problem in geometry has seemed to him of less importance than a knowledge of what the problem suggests that is stimulating to the human mind, how it is made more general, how it would appear if transformed by the shifting of lines and points, how it appears in nature and in art, where it touches human needs, and how it gives play to the imagination.

Perhaps the best description of Smith's philosophy of mathematics education is given in an excerpt from his 1921 *Religio Mathematici* to the Mathematical Association of America:

> Mathematics increases the faith of a man who has faith; it shows him his finite nature with respect to the Infinite; it puts him in touch with immortality in the form of mathematical laws that are eternal; and it shows him the futility of setting up his childish arrogance of disbelief in that which he cannot see. . . . We should teach "the science venerable" not merely for its technique; not solely for this little group of laws or that; not only for a body of unrelated propositions or for some examination set by the schools; but we should teach it primarily for the beauty of the discipline, for "the music of the spheres," and for the faith it gives in truth, in eternal law, in the Infinite, and in the reality of the imaginary; and for the feeling of humility that results from our comparison of the laws within our reach

> and those which obtain in the transfinite domain. With such a spirit to guide us, what teachers we would be!

This philosophy directly influenced several generations of mathematics educators and indirectly continues to influence students of mathematics and its teaching through the extant writings of Smith. David Eugene Smith died July 29, 1944.

References

"David Eugene Smith—A Sketch", *Scripta Mathematica* (1945) 11: 364–69.

Eisele, Carolyn and Boyd, Lyle G. "Smith, David Eugene", *Dictionary of American Biography,* Supplement 3 (1941–45). New York: Charles Scribner and Sons, 1973, 721–722.

Frick, Bertha M. "Bibliography of the Historical Writings of David Eugene Smith", *Osiris* (January 1936) 1: 9–78.

Roger, Joe T. "The Philosophy of Mathematics Education Reflected in the Life and Works of David Eugene Smith", *Peabody Contribution to Education no. 1171.* Ph.D. dissertation, George Peabody College for Teachers, 1976.

Bibliography of Historical References

Adelard of Bath (1120). *Regulae abaci.*

Alexandre de Villedieu (c. 1240). *De computo ecclesiastico.*

———*De arte numerandi.*

———*Carmen de algorism.*

Alcuin of York (c. 775). *Propositiones ad acuendos juvenes.*

al-Khwarizmi [Abu Jafar Muhammed ibn Masa al-Khwarizmi]. *Kitabhisab al-dad al-hindi.* Bagdad, c. 830.

Baker, Humphrey. *The Well Spring of Sciences.* London, 1568.

Bede the Venerable (c. 673–735). *De temporum.* 726.

——— *De numeris.*

——— *De numerorum divisione.*

Beha Eddin al-Amili Mohammed ibn Hosun. *Kholasat al-hisab.* 1600.

Benedetto da Firenzi. *Inchominica el trattato darisme tricha espelialmete quella pte che e sotto posta mercatatia e comminciando alnome didio.* Florence, c. 1460.

Bhaskara Achabrya. *Lilavati.* c. 1150.

de Boissiere, Claude, *L'art D'arythmetique.* Paris, 1554.

de Beldamandi, Prosdocemo. *Algorithmus de integris.* 1410.

Boethius, Anicius Manlius Severinus (c. 475–524). *Arithmetica Boeij.* Augsburg, 1488.

Borghetti, Smiraldo. *Opera d'abbaco.* Venice, 1594.

Borghi, Pietro. *Qui comenza la nobel opera de arithmetica. . . .* Venice, 1484.
Calandri, Philippi. *Pictagores arithmetrice introductor.* Florence, 1491.
Cardan, Jerome [Hieronymus Cardanus]. *Practica Arithmetice.* Milan, 1539.
——— *Ars Magna.* Nürnberg, 1545.
Cardinael, Sybrandt Hansz. *Arithmetica.* Amsterdam, 1659.
Cassidorius, Magnus Aurelius (c. 470–564). *De artibus ac disciplinis liberalium literarum Computus Paschalis sive de indicationibus cyclis solis et lunae.*
Cataneo, Pietro. *Le pratiche delle due prime mathematiche.* Venice, 1546.
Cattaldi [Cataldo], Pietro Antonio. *Due Lettioni.* Bologna, 1577.
Champenois, Jacques Chauvet. *Les institutions de l'arithmetique.* Paris, 1578.
Chiu chang suan shu. c. 200 B.C.–c. 200 A.D.
Chuquet, Nicolas. *Le triparty en la science des nombres.* Lyons, 1484.
Ciruelo, Pedro Sanchez. *Tractatus arithmetice practice qui dicitur algorismus.* Paris, 1513.
Clavius, Christopher. *Epitome arithmeticae practica.* Rome, 1583.
Clichtoveus, Jodocus. *Introductio in libros arithmeticos diui Seuerin Boeij. . . .* Paris, 1503.
Crafte of Nombrynge. 1300.
Digges, Leonard and Thomas. *An Arithmeticall Militare Treatise named Strativticos.* London, 1572.
Diophantus (c. 275). *Arithmetica.*
Feliciano da Lazesio, Francesco. *Libro di arithmetica et geometria.* Venice, 1526.
Forestani, Lorenzo. *Pratica d'arithmetica e geometria. . . .* Venice, 1603.
Frisius, Gemma. *Arithmeticae practicae methodus facilis.* Antwerp, 1540.
George of Hungary. *Arithmeticae summa tripartita.* 1499.
Gerbert of Aurillac. *Regulae de numerorum abaci rationibus.* c. 1003.
Ghaligai, Francesco. *Practica d'arithmetica.* Florence, 1521.
Gerardi, Paolo. *Libro di raginoni.* 1328.
Grammateus, Henricus. *Rechenbüchlin aff alle Kauffmanschafft.* Frankfort, 1535.

Heere, Johann. *Rechenbüchlein von allerhand gebraüchlichen Fragen*. Nürnberg, 1599.
Helmreich, Andreas. *Rechenbuch*. Eisleben, 1561.
Hodder, James. *Arithmetick or that Necessary Art Made Most Easy*. London, 1672.
Huswirt, Johann. *Enchiridion algorismi*. . . . Cologne, 1501.
Hylles, Thomas. *The Arte of Vulgar Arithmeticke both in Integers and Fractions*. London, 1600.
Isidorus of Seville (c. 610). *Etymologies*.
Johannaes Hispalensis (c. 1140). *Liber abaci*.
——— *Liber algorismi*.
Jordanus Nemorarius. *Arithmetica decem libris demonstrata*. Paris, 1496.
Köbel, Jacob. *Rechenbüchlein*. Augsburg, 1514.
Leonardo of Pisa [Fibonacci]. *Liber abaci*. Pisa, 1202.
Maurolycus, Franciscus. *Arithmeticorum libri duo*. Venice, 1575.
Maximus Planudes (c. 1340). Ψηο φρ'ια κα τ' ΄Τυδ'υs. [Indian Arithmetic].
Napier, John. *Mirifici logarithmorum canonis descriptio*. . . . Edinburgh, 1614.
Nicolas, Gaspar. *Tratado da pratica d'arismetica*. Lisbon, 1519.
Nicomachus of Gerasa (c. 100). *Introductionis arithmeticae libri duo*.
Noviomagus, Johann. *De numeris libri II*. Cologne, 1544.
de Ortega, Juan. *Suma de arithmetica*. Rome, 1512.
Oughtred, William. *Clavis mathematicae*. London, 1631.
Pacioli [Paciuolo], Luca. *Suma de arithmetica geometria proportioni et proportionalita*. Venice, 1494.
Pagnini, Guglielmo. *Practica mercantile moderna*. Lucca, 1562.
Pellos, [Pellizzati], Francesco. *De la art arithmeticha compendio de lo abaco*. Turin, 1492.
Petzensteiner, Heinrich. *The Bamberg Arithmetic*. Bamberg, 1483. Since the author and title remain unknown, the book is identified with the place of its origin and its printer, Petzensteiner.
Rabbi ben Ezra (c. 1140). *Sefer ha-Mispar*.
Raets, Willem. *Arithmetica*. Antwerp, 1580.
Rahn, Johann Heinrich. *Teutsche Algebra*. Zürich, 1659.
Ralph of Laon (c. 1150). *Liber de abaco*.
Ramus, Petrus. *Arithmeticae libri tres*. Paris, 1555.

Recorde, Robert, *The Ground of Artes*. London, 1542.
Riese, Adam. *Rechnung auff der Linien und Federn*. Erfurt, 1518.
Robert of Chester (c. 1140). *Algoritmi de numero indorum*.
Rudolff, Christoff. *Kunstliche rechnung mit ziffer unnd mit den zal pfenninge sampt der Wellischen Practics*. Nurnberg, 1534.
———*Exempel Büchlin*. Augsburg, 1530.
Sacrobosco [Johannes de Sacrobosco]. *Algorismis vulgaris*. 1230.
Sanct Climent, Francesch. *Suma de la art de arismetrica*. Barcelona, 1482.
Savonne, Pierre. *L'Arithmetique*. Lyons, 1563.
Scheubel, Johann. *De numeris et diversis rationibus*. Leipzig, 1545.
Sfortunati, Giovanni. *Libro di arithmetica*. Venice, 1544.
Stevin, Simon. *L'Arithmetique*. Leyden, 1585. Part three of this work, *La practique d'arithmetique*, contains a section devoted to *La disme* [De thiende].
Stifel, Michael. *Deutsche Arithmetica*. Nürnberg, 1545.
———*Die Coss*. Königsberg, 1553–54.
Tagliente, Girolamo and Giannantionio. *Libro dabaco*. Venice, 1515.
Tartaglia, Nicolo. *La prima parte del general trattato di numeri, et misure*. Venice, 1556.
Tonstall, Cuthbert. *De arte supputandi libri quattuor*. London, 1522.
Trechant, Ian. *L'Arithmetique*. Lyon, 1578.
Turchillus. *Regunculi super abacum*. c. 1200.
Van der Schuere, Jacob. *Arithmetica*. Haarlem, 1600.
de Versi, Piero. *Alcune raxion dei marineri*. Venice, 1444.
Wagner, Ulrich. *The Bamberg Arithmetic*. Bamberg, 1482. This is the first printed German arithmetic. Its original title remains unknown and it is thus referred to as either the *Bamberg Arithmetic* or *Petzensteiner's Arithmetic* after its printer, Heinrich Petzensteiner.
Widman, Johann. *Behede und hubsche Rechnung auf allen kauffmanschafft*. Leipzig, 1489.

General Bibliography

Al-Daffa, Abdullah. *The Muslim Contributions to Mathematics*. London: Croom Helm, Ltd., 1977.

Arrighi, Gino. "A proposito di 'abachisti' e 'algoritmisti'," *Physis* (1973) 15: 107–110.

———"Un 'programma' di didattica di matematica della prima meta del Quattrocento," *Atti e memorie della' Academic Petrarca di lettere, arti e Scienze* (1968) 38: 117–128.

———"Les mathematiques italiennes vers la fin du moyen-age et à l'epoque de la Renaissance," *Organon* (1968) 5: 181–188.

———"La matematica del Rinascimento in Firenze. L'eredita di Leonardo Pisano e le Botteghe d'Abaco'," *Cultura e scuola* (1966) 5: 287–294.

———"L'aritmetica' di Paolo dell'Abbaco e i tempi sui," *Cultura e scuola* (1966) 5: 287–294.

———"Regole d'abaco nei primi secoli dei numeri in 'figure degli Indi'," *Bollettino della Unione matematica italiana* (1964) 19: 490–502.

Benedict, Suzan Rose. *A Comparative Study of Early Treatises Introducing into Europe the Hindu Art of Reckoning*. Concord, N.H.: Rumford Press, 1914.

Boncompagni, Baldassarre. "Intorno ad un trattato d'arithmetica stampato nel 1478," *Atti dell' Accademia Pontifica de' Nuovi Lincei*, 16 (1862–63) 1–61.

Boorstin, Daniel J. *The Discoverers.* New York: Random House, 1983.

Boyer, Carl B. *A History of Mathematics.* New York: John Wiley & Sons, Inc., 1968.

Bronowski, Jacob. *Magic, Science and Civilization.* New York: Columbia University Press, 1978.

Brown, Horatio F. *The Venetian Printing Press.* New York: G. P. Putnam's Sons, 1891.

Brown, R. G. *Pacioli on Accounting.* New York: McGraw-Hill, 1963.

Brunet, J. C. *Manuel der libraire et de l'ameteur de livres.* Paris, 1862.

Burke, Peter. *Venice and Amsterdam: A Study of Seventeenth Century Elites.* London: Temple Smith, 1974.

Cajori, Florian. *A History of Mathematical Notation.* 2 vols. La Salle, Ill: Open Court Publishing Co., 1928.

Campbell, Douglas M. *The Whole Craft of Number.* Boston: Prindle, Weber & Schmidt, Inc., 1976.

Chace, A. B., et al. *The Rhind Mathematical Papyrus.* 2 vols. Buffalo, N.Y.: Mathematical Association of America, 1927–29.

Chambers, D. S. *The Imperial Age of Venice, 1380–1580.* Baltimore: Johns Hopkins University Press, 1973.

Cipolla, Carlo M. *Money, Prices and Civilization in the Mediterranean World.* New York: Gordian Press, Inc., 1967.

Coghill, Nevill. *The Canterbury Tales.* Harmondsworth: Penguin Books, 1951.

Colebrooke, H. T. *Algebra with Arithmetic and Mensuration from the Sanskrit of Brahmagupta and Bhascara.* Calcutta, 1927.

Cowley, Elizabeth Buchanan. "An Italian Mathematical Manuscript," *Vassar Medieval Studies.* New Haven: Yale University Press, 1923, 379–405.

Dahl. Svend. *History of the Book.* Metuchen, N.J.: Scarecrow Press, 1968.

d'Andeli, Henri. *La Bataille des VII Arts.* Berkeley: University of California Press, 1927.

Dantzig, Tobias. *Number: The Language of Science.* New York: Macmillan & Co., 1954.

Davis, Charles. "Education in Dante's Florence," *Speculum* (1965) 40: 415–435.

Davis, Margaret Daly. *Piero della Francesca's Mathematical Treatises*. Revenna: Longo, 1977.
Davis, Natalie Zemon. "Sixteenth Century French Arithmetics on the Business Life," *Journal of the History of Ideas* (1960) 21: 18–48.
Dawson, Warren. *A Leechbook or Collection of Medical Recipes of the Fifteenth Century*. London: Macmillan & Co., 1934.
DeMorgan, Augustus. *Arithmetical Books from the Invention of Printing to the Present Time*. London: Taylor and Walton, 1847.
DeRoover, Raymond. "The Development of Accounting prior to Luca Pacioli according to the Account Books of Medieval Merchants," *Business, Banking and Economic Thought in Late Medieval and Early Modern Europe*. Julius Kirshner (ed.) Chicago: University of Chicago Press, 1974.
———"A Florentine Firm of Cloth Manufacturers," *Speculum* (1941) 16: 3–33.
Edler, Florence. *Glossary of Medieval Terms of Business*. Cambridge, Mass.: The Medieval Academy of America, 1934.
Eisenberg, Murry. *Axiomatic Theory of Sets and Classes*. New York: Holt Rinehart and Winston, Inc., 1971.
Enciclopedia italiana. Milan: Rizzoli and Co., 1937.
Evans, Gillian R. "From Abacus to Algorism: Theory and Practice in Medieval Arithmetic," *The British Journal for the History of Science* (1977) 10: 114–131.
Fanfani, Amintore. "La preparation intellectuelle et professionnelle a l'activité economique en Italie du xive et xvie siecle," *Le Moyen Age* (1951) 57: 327–346.
Frederici, Domenico Maria. *Memorie Trevigiane sulla Tipopgrafia del secolo XV*. Venice, 1805.
Ferguson, W. K. "Recent Trends in the Economic Historigraphy of the Renaissance," *Studies in the Renaissance* (1960) 7: 19–26.
Frick, Bertha M. "The First Portuguese Arithmetic," *Scripta Mathematica* (1943) 11: 327–339.
Gail, Marzieh. *Life in the Renaissance*. New York: Random House, 1968.
Gardner, Arthur O. "The History of Mathematics as a Part of the History of Mankind," *The Mathematics Teacher* (1968) 61: 524–526.

Gies, Joseph and Gies, Francis. *Leonardo of Pisa and the New Mathematics of the Middle Ages*. New York: Thomas Y. Crowell Co., 1969.

Glaisher, J. W. L. "On the Early History of the Signs + and − and on the Early German Arithmeticians," *Messenger of Mathematics* (1921–22) 15: 1–148.

Goldthwaite, Richard A. "Schools and Teachers of Commercial Arithmetic in Renaissance Florence," *Journal of Economic History* (1972) 1: 418–433.

Grand, Edward (ed.). *A Source Book in Medieval Science*. Cambridge, Mass.: Harvard University Press, 1974.

Graesser, R. F. "The Date of Easter," *School Science and Mathematics* (1952) 52: 371–372.

Groseclose, Elgin. *Money and Man: A Survey of Monetary Experience*. New York: Fredrick Ungar Publishing Co., 1961.

Grosse, Hugo. *Historische Rechenbücher des 16 and 17 Jahrhundertes und die Entwicklung ihrer Grundgedanken bis zur Neuzeit*. Leipzig: Dürr, 1901.

Hale, J. R. *Renaissance Europe: Individual and Society 1480–1520*. Berkeley: University of California Press, 1971.

Hill, G. F. *The Development of Arabic Numerals in Europe*. Oxford: Clarendon Press, 1915.

Hirsch, S. Carl. *Meter Means Measure*. New York: Viking Press, 1973.

Hoffmann, Joseph E. *The History of Mathematics*. New York: The Philosophical Library, 1957.

Ives, Herbert E. *The Venetian Gold Ducat and its Imitations*. New York: The American Numismatic Society, 1954.

Jackson, Lambert L. *The Educational Significance of Sixteenth Century Arithmetic*. New York: AMS Press, 1972. Reprint 1906 edition.

Jeannin, Pierre. *Merchants of the 16th Century*. New York: Harper and Row Publishers, 1972.

Karpinski, L. C. "The Italian Arithmetic and Algebra of Master Jacob of Florence," *Archeion* (1929) 11: 170–177.

———"An Early Printed Italian Arithmetical Treatise," *Archeion* (1929) 11: 331–335.

———"The First Printed Arithmetic of Spain, Francesch Sanct Climent: Suma de la art arismetrica Barcelona 1482," *Osiris* (1936) 1: 411–420.

———*The History of Arithmetic*. New York: Russell & Russell, Inc., 1965.

La libreria gia raccolta con grande studio dal Signor Maffeo Pinelli Veneziano. 4 vols. Venice, 1787.

Lam Lay Yong. "On the Chinese Origin of the Galley Method of Arithmetical Division," *The British Journal of the History of Science* (1966) 3: 66–69.

Lane, Frederic. *Venice and History*. Baltimore: Johns Hopkins University Press, 1966.

———*Andrea Barbarigo, Merchant of Venice 1418–1449*. New York: Octagon Books, 1967.

———*Venice: A Maritime Republic*. Baltimore: Johns Hopkins University Press, 1973.

Lopez, Robert. "The Trade of Medieval Europe: The South," *The Cambridge Economic History of Europe*. M. Postan and E. E. Rich (eds.) Cambridge: Cambridge University Press, 1952, 2: 257–338.

———"Hard Times and Investment in Culture," *The Renaissance: A Symposium*. New York: The Metropolitan Museum of Art, 1953.

———"Stars and Spices: The Earliest Italian Manual of Commercial Practice," *Economy, Society and Government in Medieval Italy*. Davis Herlihy, Robert Lopez and Vsevolvd Slessarev (eds.) Kent, Ohio: Kent State University Press, 1969, 35–42.

———*The Commercial Revolution of the Middle Ages*. Englewood Cliffs, N.J.: Prentice-Hall, 1971.

Luzzatto, Gino. *An Economic History of Italy*. London: Routledge and Kegan Paul, 1961.

McGovern, John F. "The Rise of New Economic Attitudes—Economic Humanism, Economic Nationalism—During the Later Middle Ages and the Renaissance, A.D. 1200–1550," *Traditio* (1970) 26: 215–253.

McMurtrie, Douglas C. *The Book: The Story of Printing and Bookmaking*. New York: Oxford University Press, 1967.

Mendelssohn, Kurt. *Science and Western Domination*. London: Thames and Hudson, 1976.

Menniger, Karl. *Number Words and Number Symbols*. Cambridge, Mass.: MIT Press, 1969.

Miller, G. A. "On the History of Common Fractions," *School Science and Mathematics* (1931) 31: 138–145.

Miskimin, Harry A. *The Economy of Early Renaissance Europe, 1300–1460*. Englewood Cliffs, N.J.: Prentice-Hall, 1969.

Molment, Pompeo. *Venice: Its Individual Growth from the Earliest Beginnings to the Fall of the Republic*. 2 vols. Chicago: A. C. McClurg & Co., 1906.

National Geographical Society. *The Renaissance: Maker of Modern Man*. Washington, D.C.: The National Geographical Society, 1970.

Nelson, Benjamin. "The Usurer and the Merchant Prince: Italian Businessmen and the Ecclesiastical Law of Restitution, 1100–1550," *Journal of Economic History* (1947) 7 (Sup.): 104–22; John F. McGovern.

———*The Idea of Usury*. Princeton: Princeton University Press, 1949.

Neugebauer, O. *The Exact Sciences in Antiquity*. New York: Dover Publications, 1969.

Oates, J. C. T. (ed.). *A Catalogue of the Fifteenth-Century Printed Books in the University Library Cambridge*. Cambridge: Cambridge University Press, 1954.

Parry, John W. *The Story of Spices*. New York: Chemical Publishing Co., 1953.

Payen, John (translator). *The Decameron of Giovanni Boccaccio*. New York: Horace Liveright, 1925.

Peragallo, Edward. *Origin and Evolution of Double Entry Bookkeeping: A Study of Italian Practice Since the Fourteenth Century*. New York: American Institute Publishing Co., 1938.

Perkins, Samuel, "The Magic of Saffron," *Silver Kris* (1979) 5: 29–31.

Pullan, Brian. *A History of Early Renaissance Italy*. New York: St. Martin's Press, 1973.

———*Rich and Poor in Renaissance Venice*. Oxford: Basil Blackwell, 1971.

Pullan, J. M. *The History of the Abacus*. New York: Frederick A. Praeger, 1969.

Reynolds, Robert L. *Europe Emerges*. Madison, Wisc.: University of Wisconsin Press, 1961.

Riccardi, P. *Biblioteca Mathematica Italiano*. Modena, 1893.

Rogers, J. T. *The Philosophy of Mathematics Education Reflected in the Life and Work of David Eugene Smith*. Unpublished doctoral dissertation. Peabody College for Teachers, 1976.

Rose, Paul L. *The Italian Renaissance of Mathematics.* Geneva: Librairie Droz, 1975.

Sanford, Vera. "The Art of Reckoning," *The Mathematics Teacher* (1951) 44: 135–137.

———*The History and Significance of Certain Standard Problems in Algebra.* Reprint of 1927 edition. New York: AMS Press, 1972.

Sarton, George. *Introduction to the History of Science.* 2 vols. Baltimore: The Williams and Wilkins Co., 1927.

———"Arabic Commercial Arithmetic," *Isis* (1933) 20: 260–262.

———"Science in the Renaissance," *The Civilization of the Renaissance.* James W. Thompson et al. (eds.) Chicago: University of Chicago Press, 1959.

———*Appreciation of Ancient and Medieval Science During the Renaissance.* New York: A. S. Barnes and Company, 1961.

Schrader, Dorothy. "The Arithmetic of the Medieval Universities," *The Mathematics Teacher* (1968) 61: 615–628.

———"*De Arithmetica,* Book I of Boethius," *The Mathematics Teacher* (1968) 61: 615–628.

Selfridge, H. Gordon. *The Romance of Commerce.* London: John Lane, The Bodley Head, Ltd., 1923.

Shirk, J. A. G. "Contributions of Commerce to Mathematics," *The Mathematics Teacher* (1939) 32: 203–208.

Simons, L. G. "Historical Notes on Arithmetical Division," *Scripta Mathematica* (1932) 1: 362–363.

Sleight, E. R. "The Craft of Nombrynge," *The Mathematics Teacher* (1939) 32: 243–248.

Smith, D. E. "The Treviso Arithmetic, 1478; *(Arte de'ell'abbaco,* Groff A 114–1)." Unpublished manuscript and notes (1907–1911). New York: The David Eugene Smith Collection, Columbia University Library.

———*Rara Arithmetica.* Boston: Ginn & Co., 1908.

———"Medicine and Mathematics in the Sixteenth Century," *Annals of Medical History* (1917) 1: 125–139.

———"On the Origin of Certain Typical Problems," *American Mathemathical Monthly* (1917) 24: 64–71.

———"Mathematical Problems in Relation to the History of Economics and Commerce," *American Mathematical Monthly* (1918) 24: 221–223.

———"The First Printed Arithmetic," *Isis* (1924) 6: 310–31.

———*A Source Book in Mathematics*. New York: McGraw-Hill, 1929.

———The Influence of the Mathematical Works of the Fifteenth Century upon those of Later Times," *Papers of the Bibliographical Society of America* (1932) 26: 143–171.

———*History of Mathematics*. 2 vols. New York: Dover Publications, 1958.

Solomon, Bernard S. "One is No Number in China and the West," *Harvard Journal of Asiatic Studies* (1954) 17: 253–260.

Spahr, Rodolfo. *Le Monete Siciliane*. Zürich: Graz, 1976.

Steele, R. "What Fifteenth Century Books are About," *The Library* (1903) 4: 337–354.

Steele, Robert. *The Earliest Arithmetic in English*. London: Early English Text Society, 1922.

Stillwell, Margaret Bingham. *The Awakening Interest in Science during the First Century of Printing 1450–1550*. New York: The Bibliographical Society of America, 1970.

Struik, Dirk J. *A Concise History of Mathematics*. New York: Dover Publication, Inc., 1967.

———"The Prohibition of the Use of Arabic Numerals in Florence," *Archives internationale d'historire de sciences* (1968) 21: 291–294.

Sullivan, J. W. N. *The History of Mathematics in Europe*. London: Oxford University Press, 1925.

Swetz, F., and Kao, T. I. *Was Pythagoras Chinese?* University Park, Penn.: Pennsylvania State University Press, 1977.

Symonds, John Addington. *Renaissance in Italy: The Revival of Learning*. Gloucester, Mass.: Peter Smith, 1967.

Tannery, Paul. *Notices et extraits des manuscrits de la Bibliotheque nationale* 32: 167.

Tawney, R. H. *Religion and the Rise of Capitalism*. New York: Harcourt, Brace and Company, 1926.

Thompson, James W. *The Literacy of the Laity in the Middle Ages*. New York: Burt Franklin, 1960.

Thorndike, Lynn. *Science and Thought in the Fifteenth Century*. New York: Columbia University Press, 1929.

———"Renaissance or Prenaissance?," *Journal of the History of Ideas* (1943) 4: 65–74.

Usher, Abblot Payson. *The Early History of Deposit Banking in Mediterranean Europe*. New York: Russell & Russell, 1943.

Van Egmond, Warren. *The Commercial Revolution and the Beginnings of Western Mathematics in Renaissance Florence, 1300–1500*. Ph.D. dissertation, History of Science, Indiana University. Ann Arbor: University Microfilm International, 1976.

—*Practical Mathematics in the Italian Renaissance: A Catalogue of Italian Abbacus Manuscripts and Printed Books to 1600*. Florence: Instituto e Museo di Storia della Scienza Firenze, 1980.

Vogel, Kurt. *Die Practica des Algorismus Ratisbonensis*. Munich: C. H. Beck'sche Verlagsbuchhandlung, 1954.

———*"Algorismus", das früheste Lehrbuch zum Rechner mit indeschen Ziffern nach der unzigen lateinischen Handschrift (Camb. Univ. Lib. MSI: 6, 5)* Aalen: O. Zeller, 1963.

———"Fibonacci, Leonardo or Leonardo of Pisa," *Dictionary of Scientific Biography*. New York: Scribner Publishing, 1971, 4: 604–613.

———*Ein Italienisches Rechenbuch aus dem 14 Jahrhundert*. Munich: Deutschen Museums für die Geschichte der Naturwissenschaften und der Technik, 1977.

Wang, Ling. "The Chinese Origin of the Decimal Place-Value System in the Notation of Numbers," Communication to the 23rd International Congress of Orientalists, Cambridge, England, 1954.

Water, E. G. R. "A Thirteenth Century Algorism in French Verse," *Isis* (1920) 11: 45–82.

———A Fifteenth Century French Algorism from Liege," *Isis* (1931) 22: 192–236.

Weissenborn, Gerbert H. *Beitrage zur Kenntniss der Mathematik des Mittelalters*. Berlin, 1888.

Wilson, P. W. *Romance of the Calendar*. New York: W. W. Norton, 1937.

Witz, P. J. "Dixit algorismi: An annotated translation with commentary." Unpublished student paper,

Department of Mathematical Sciences, University of Southampton, 1980.

Wolf, A. *A History of Science, Technology and Philosophy in the Sixteenth and Seventeenth Centuries*. Vol. 1. New York: Harper and Brothers, 1959.

Woodward, William H. *Studies in Education during the Age of the Renaissance 1400–1600*. New York: Russell & Russell, 1965.

Wussing, Hans L. "European Mathematics During the Evolutionary Period of Early Capitalistic Conditions," *Organon* (1967) 4: 89–93.

Yeldham, Florence A. *The Story of Reckoning in the Middle Ages*. London: George G. Harrap & Co., 1926.

Youschkevitch, A. P. *Geschichte der Mathematik in Mittelalter*. Leipzig: Teubner, 1964.

Zinberg, George. *Jewish Calendar Mystery Dispelled*. New York: Vantage Press, 1963.

Zupko, Ronald Edward. *A Dictionary of English Weights and Measures*. Madison, Wisc.: University of Wisconsin Press, 1968.

Notes

Notes to Chapter 1

1. John F. McGovern, "The Rise of New Economic Attitudes—Economic Humanism, Economic Nationalism—During the Later Middle Ages and the Renaissance, A.D. 1200–1550", *Traditio* (1970), 26:215–253.

2. The previous historical period when such humanistic circumspection took place was during the Hellenic era of western civilization (c. 400 B.C.–30 B.C.).

3. Specific dating for this period varies with authorities and topics of discussion. For example, Robert Lopez, in speaking of the Renaissance in broad terms as a European intellectual movement, sets its date as 1330–1530; George Sarton dates the Scientific Renaissance 1450–1600, Marie Boas dates the same phenomena 1450–1630; one can also speak of the Italian Renaissance (1420–1550) or the Renaissance of Arts and Letters (1400–1600). In such discussions, boundary dates for the period are relative and extended or contracted to include or exclude certain historical events or personages. The word "Renaissance" evolved from the French words "renaître", to be born again, and "naissance", birth. It

was first used to classify the historical era spanning the fourteenth to the seventeenth centuries by the French historian Jules Michelet, who in 1855 entitled the seventh volume of his *History of France, The Renaissance.*

4. Venice's identity can be traced back to 568 A.D. when the Lombard invasion of Italy forced mainland refugees to flee to the mud banks and salt marshes along the northern Adriatic Coast. In this hostile environment they established a flourishing civilization.

5. Quoted in D. S. Chambers, *The Imperial Age of Venice, 1380–1580* (London: Thames and Hudson, Ltd., 1970).

6. The power and extent of the Venetian Empire is described in Frederic C. Lane, *Venice: A Maritime Republic* (Baltimore: The Johns Hopkins University Press, 1973).

7. Ibid., p. 21.

8. For a discussion of the Venetian wool industry see Robert Lopez, "The trade of Medieval Europe: The South", *The Cambridge Economic History of Europe,* vol. 2, edited by M. Postan and E. E. Rich (Cambridge: The University Press, 1952), pp. 257–338.

9. Lane, op. cit., p. 61.

10. *See* Treviso, *Enciclopedia Italiana,* vol. 34 (Milano: Rizzoli and Company, 1937), pp. 284–88.

11. Brian Pullan, *Rich and Poor in Renaissance Venice* (Oxford: Basil Blackwell, 1971), p. 27.

12. Peter Burke, *Venice and Amsterdam: A Study of Seventeenth-Century Elites* (London: Temple Smith, 1974), p. 49.

13. Brian Pullan, *A History of Early Renaissance Italy* (New York: St. Martin's Press, 1973), p. 108.

14. Gino Luzzatto, *An Economic History of Italy* (London: Routledge and Kegan Paul, 1961), p. 28.

15. Warren Van Egmond, "The Commercial Revolution and the Beginnings of Western Mathematics in Renaissance Florence, 1300–1500" (Ann Arbor: University Microfilms International, 1976), p. 136. Ph.D. dissertation, Department of History and Philosophy of Science, Indiana University.

16. An extensive listing of these terms is given in Florence Edler, *Glossary of Medieval Terms of Business* (Cambridge, Mass.: The Medieval Academy of America, 1934).

17. Pierre Jeannin, *Merchants of the 16th Century* (New York: Harper and Row Publishers, 1972), p. 16.

18. *See* Joseph Gies and Frances Gies, *Leonardo of Pisa and the New Mathematics of the Middle Ages* (New York: Thomas Y. Crowell Co., 1969).

19. For a complete discussion of Fibonacci and his work, *see* the entry by Kurt Vogel, "Fibonacci, Leonardo or Leonardo of Pisa", *Dictionary of Scientific Biography* (New York: Scribner's, 1971) 4:604–613.

20. Raymond de Roover, "The Development of Accounting prior to Luca Pacioli according to the Account Books of Medieval Merchants", *Business Banking and Economic Thought in Late Medieval and Early Modern Europe,* Julius Kirshner (ed.) (Chicago: The University of Chicago Press, 1974), pp. 119–180.

21. Gino Luzzatto, *An Economic History of Italy from the Fall of the Roman Empire to the Beginning of the Sixteenth Century* (London: Routledge and Kegan Paul, 1961), p. 118; *see also* K. S. Johnston and R. G. Brown, *Pacioli on Accounting* (New York: McGraw Hill, 1963), and Edward Peragallo, *Origin and Evolution of Double Entry Bookkeeping: A Study of Italian Practice Since the Fourteenth Century* (New York: American Institute Publishing Company, 1938).

22. Among European bookkeepers well into the 18th century the terms 'after the Italian manner' and 'double entry' were synonymous.

23. Pompeo Molment, *Venice: Its Individual Growth from the Earliest Beginnings to the Fall of the Republic,* 2 vols. (Chicago: A. C. McClurg and Co., 1906) 1:141.

24. Tobias Dantzig, *Number: The Language of Science* (New York: Macmillan Co., 1954), p. 26.

25. For a discussion of the mathematics of medieval universities, *see* Dorothy Schrader, "The Arithmetic of the Medieval Universities", *The Mathematics Teacher* (1967) 60:264–74.

26. Henri d'Andeli, *La Bataille des VII Arts* (Berkeley: The University of California Press, 1927), pp. 48–49.

27. For a complete discussion of the reckoning master's position in his community, *see* Van Egmond, op. cit., Chapter 3 and Appendix A.

28. Karl Menninger, *Number Words and Number Symbols* (Cambridge, Mass.: The MIT Press, 1969), p. 435.

29. Ibid., p. 428.

30. Lynn Thorndike, "Elementary and Secondary Education in the Middle Ages", *Speculum* (1940) 15:400–408.

31. James Westfall Thompson, *The Literacy of the Laity in the Middle Ages* (New York: Burt Franklin, 1960), p. 53; Charles T. Davis, "Education in Dante's Florence", *Speculum* (1965) 40:415–435.

32. For a detailed discussion of the growth of the merchant trade, *see* Pierre Jeannin, op. cit.

33. Jeannin, op. cit., p. 82.

34. Amintore Fanfani, "La préparation intellectuelle et professionnelle à l'activité économique, en Italie du xive and xvie siècle", *Le Moyen Age* (1951) 57:327–346.

35. Gino Arrighi, "Un'Programma'di Didattica di Matematica nella Prima metà del Quattrocento", *Atti e memorie della Academia Petrarca di lettere, arti e scienze* (1968) 38:117–128.

36. Richard A. Goldthwaite, "Schools and Teachers of Commercial Arithmetic in Renaissance Florence", *Journal of European Economic History* (1972) 1:418–433.

37. The *Treviso Arithmetic* is the earliest dated European arithmetic in existence. Claims have been made as to the prior publication of other arithmetics; however, the publication dates of these books are unknown and must be approximated from style or type-font used. Karpinski describes an anonymous arithmetic published in Venice by Adam de Rottweil in "An Early Printed Italian Arithmetical Treatise", *Archeion* (1929) 11:331–35, and estimates its date as 1476/78, but possibly as late as 1480. Stillwell, op. cit, lists *Algorithms* (Trent: Albrecht Kunne, c. 1475, p. 41). For a survey of the contents of fifteenth-century books, *see* R. Steele, "What Fifteenth Century Books are About", *The Library* (1903) 4:337–54.

38. Augustus de Morgan, *Arithmetical Books from the Invention of Printing to the Present Time* (London: Taylor and Walton, 1847), p. 1.

39. For a discussion of early printed arithmetics, particularly those of Portugal, *see* Bertha M. Frick, "The First Portuguese Arithmetic", *Scripta Mathematica* (1943), 11:327–339; L. C. Karpinski, "The First Printed Arithmetic of Spain, Francesch Sanct Climent: *Suma de la art de arismetrica,* Barcelona 1482", *Osiris* (1936) 1:411–20.

40. Douglas C. McMurtrie, *The Book: The Story of Printing and Bookmaking* (New York: Oxford University Press, 1967), p. 283.

41. Frick, op. cit., p. 333.

42. McMurtrie, op. cit., indicates typographical difficulties, p. 283.

43. Horatio F. Brown, *The Venetian Printing Press* (New York: G. P. Putman's Sons, 1891), p. 2.

44. McMurtrie, op. cit., p. 214.

45. Domenico Maria Federici, *Memorie Trevigiane sulla Tipopgrafia del secolo XV* (Venice, 1805), p. 73; Margaret Bingham Stillwell, *The Awakening Interest in Science during the First Century of Printing 1450–1550* (New York: The Bibliographical Society of America, 1970).

46. *See* Dorothy V. Schrader, "De Arithmetica, Book I of Boethius", *The Mathematics Teacher* (1968) 61:615–628.

47. For a discussion of "al-Khwarizmi" and his works, *see* Ali Abdullah al-Daffa, *The Muslim Contributions to Mathematics* (London: Croom Helm, Ltd., 1977), pp. 49–59.

48. As translated in Menninger, op. cit., p. 411. A facsimile transcription and commentary of *Algoritmi de numero indorum* is given by Kurt Vogel, *"Algorismus", das früheste Lehrbuch zum Rechner mit indeschen Ziffern nach der unzigen lateinischen Handschrift (Camb. Univ. Lib. M.S.I.: 6, 5)* (Aalen: O. Zeller, 1963). The manuscript has been translated into English by P. J. Witz, "Dixit algorismi: An annotated translation with commentary", unpublished student paper, Department of Mathematical Sciences, University of Southampton, 1980.

49. Menninger, Ibid., p. 412.

50. Ibid., p. 412.

51. See Vera Sanford, "The Art of Reckoning" (IV Algorisms: Computing with Hindu-Arabic Numerals), *The Mathematics Teacher* (1951) 44:135–37.

52. E. R. Sleight, "The Craft of Nombrynge", *The Mathematics Teacher* (1939) 32:243–248.

53. *See,* for example, E. G. R. Water, "A Thirteenth Century Algorism in French Verse", *Isis* (1920) 11:45–82; "A Fifteenth Century French Algorism from Liege", *Isis* (1931) 22:192–236.

54. Van Egmond, op. cit., refers to such works as "abaci".

55. A detailed discussion of the evolution of the abacus is given in J. M. Pullan, *The History of the Abacus* (New York: Frederick A. Praeger, 1968).

56. Translated in Menninger, op. cit., p. 432.

57. Gino Arrighi, "A proposito di 'abachisti' e 'algoritmisti'", *Physics* (1973) 15:107–110.

58. A discussion of the mathematical transition at this time is given in Gillian R. Evans, "From Abacus to Algorism: Theory and Practice in Medieval Arithmetic", *The British Journal for the History of Science* (1977) 10:114–131.

59. Jeannin, op. cit., p. 93.

60. These dates were found by use of the 'Golden Number' and the 'Domenical Letter'. For further information on these techniques, *see* discussion on computi in D. E. Smith, *History of Mathematics,* 2 vols. (New York: Dover Publications, 1958) 2:651–52.

61. Louis Charles Karpenski, *The History of Arithmetic* (New York: Russell & Russell, Inc., 1965), p. 68. This information is contradicted by statistics given in P. Riccardi, *Biblioteca Mathematica Italiano* (Modena, 1893): Parts 2, 11 & 15 indicate that before 1500, 214 printed works of arithmetic appeared in Italy.

62. Ibid., p. 68.

63. In 1484, Piero Borghi published a practica that became a commercial standard. Like the *Treviso Arithmetic,* the book bears no formal title and is referred to merely as *Libro de Abacho,* Venice, 1484; *see* David E. Smith, "The First Great Commercial Arithmetic", *Isis* (1926) 8:41–49.

64. Other copies of the work are contained in the libraries of Harvard University and Cambridge University. Stillwell, op. cit., indicated the existence of a copy of the *Treviso Arithmetic* in the holdings of the British Museum; however, it is not listed in the Museum's catalogue of books. In 1969, a facsimile of the book was published in Treviso.

65. *La libreria già raccolta con grande studio dal Signor Maffeo Pinelli Veneziano,* 4 vols. (Venice, 1787).

66. J. C. Brunet, *Manuel du libraire et de l'ameteur de livres* (Paris, 1862). The sale number 1132 still appears on the cover of the Columbia copy.

67. The price appears in the catalogue of the sale found at the British Museum.

Notes to Chapter 2

1. David Eugene Smith, "The First Printed Arithmetic", *Isis* (1924) 6: 310–331; Smith used translated passages from his *Isis* article to describe the contents and format of the *Treviso Arithmetic* in his *A Source Book in Mathematics* (New York: McGraw Hill, 1929), pp. 1–12. This work was reprinted by Dover Publications in 1959; also cited in his *History of Mathematics,* 2nd vol. (New York: Dover Publications, 1958).

2. Smith's early formal studies were as a classicist and he was trained in Latin, Greek, and Hebrew. During his lifetime he also developed facilities in German, French, Spanish, and Italian.

3. David Eugene Smith, *Treviso* manuscript and notes.

4. In the Italian of this period there were several spellings for the word "pound": *libbra, libra,* and *lira.* The author of the *Treviso Arithmetic* uses the latter term and this is retained in the translation.

Notes to Chapter 3

1. The author's opening word is *Incommincia* ("Incommincia una practica. . .", "Here begins a practica. . . ."). There appears to be an error of spelling or typesetting, as the word should appear as "Incomincia". It is spelled correctly in the remainder of the text.

2. Fol. 1r.

3. In the early Renaissance, before a standardization of the spelling of words, many different forms for the word *abbaco* existed. The author of the *Treviso* used the form *abbacho.* Other forms common in the literature of the time were *abaco* and *abbacco.* Throughout this study, I will use the simple form *abbaco* except where directly quoting or transcribing a reference. Fibonacci's work of 1202 is commonly referred to as *Liber abaci* and is mistranslated in many texts on the history of

mathematics to mean *Book of the Abacus*. *See,* for example, Carl B. Boyer, *A History of Mathematics* (New York: John Wiley & Sons, Inc., 1968), p. 280. Actually it should be translated as the *Book of Computation*.

4. "Omnia, quae a primaeva rerum origine processerunt, ratione numerorum formata sunt, et quemadmodum sunt, sic cognosci habent: unde in universa rerum cognitione est ars numerandi cooperativa": "All things which have originated from the beginning have been formed by means of a science of number, and to whatever extent all things exist they must be so recognized, since the science of numbering has been allied with universal knowledge"; translated in Suzan Rose Benedict, *A Comparative Study of Early Treatises Introducing into Europe the Hindu Art of Reckoning* (Concord, N.H.: Rumford Press, 1914), p. 36.

5. John of Sacrobosco (Holywood), *Algorismus vulgaris,* translated and annotated by Edward Grant, "Arabic Numerals and Arithmetic Operations in the Most Popular Algorism of the Middle Ages", in *A Source Book in Medieval Science,* Edward Grant (ed.), (Cambridge, Mass.: Harvard University Press, 1974), pp. 94–101.

6. Fol. 1r.

7. Pacioli, *Suma de arithmetica geometri proportioni et proportionalita* (1523).

8. Johann Noveomagus, *De numeris libri II* (1544).

9. Gemma Frisius, *Arithmeticae practicae methodus faciles* (1575).

10. Franciscus Mavrolycus, *Arithmeticorum libri duo* (1575).

11. For example, *see* Bernard S. Solomon, " 'One is No Number' in China and the West", *Harvard Journal of Asiatic Studies* (1954), 17: 253–260.

12. Peano devised an axiomatic system for generating the natural numbers. His system depended on the concept of a successor, that is, in certain sets, any element has another unique element immediately following it, its successor. Thus the set may be thought of as an initial element and its successors. For the natural numbers, we usually think of 1 as the first number and the other numbers generated from 1 as successors, i.e., 1, 2, 3, ... $1 + 1 = 2$, $2 + 1 = 3$, thus 2 is the successor of 1, 3 is the successor of 2, etc. *See* Murray

Eisenberg, *Axiomatic Theory of Sets and Classes* (New York: Holt Rinehart and Winston, Inc., 1971), pp. 82–83.

13. Gerbert H. Weissenborn, *Beitrage zur Kenntniss der Mathematik des Mittelalters* (Berlin, 1888), p. 219.

14. *See* the discussion of this matter in Lambert Lincoln Jackson, *The Educational Significance of Sixteenth Century Arithmetic* (New York: Teachers College Press, 1906), pp. 31–32.

15. Fol. 1r.

16. Fol. 1v.

17. The popular term 'Hindu-Arabic numerals' conveys the historical impression that the numerals originated in India and were transmitted westward by Arab agents. Recent research casts some doubt on this hypothesis and moves the site of origin further eastward to China. See Wang Ling, "The Chinese Origin of the Decimal Place–Value System in the Notation of Numbers", Communication to the 23rd International Congress of Orientalists, Cambridge, England, 1954.

18. William Shakespeare, *The Winter's Tale,* act 4, sc. 2. The clown is perplexed by the problem: "How much money can be obtained for the wool from 1500 sheep (wethers) if every eleven sheep yield one tod (28 lbs.), each tod being worth a guinea?" For other Shakespearean references to counter reckoning, *see* J. M. Pullan, op. cit., pp. 110–12.

19. George Sarton, *Introduction to the History of Science,* 2 vols. (Baltimore: The Williams and Wilkins Co., 1927). Vol. II, p. 5, states that the numerals were not popular because there was no social need for them, but the question is open as to who determined social need.

20. Menninger, op. cit., p. 426. *See also* Dirk Struik, "The Prohibition of the Use of Arabic Numerals in Florence", *Archives internationales d'histoire de sciences* (1968) 21: 291–94.

21. Pullan, op. cit., p. 34.

22. Menninger, op. cit., p. 427.

23. Menninger, op. cit., p. 427.

24. Joseph E. Hofmann, *The History of Mathematics* (New York: The Philosophical Library, 1957), p. 72.

25. *See* note in Dirk J. Struik, *A Concise History of Mathematics* (New York: Dover Publications, Inc., 1948),

p. 87; and Florence Edler, *Glossary of Medieval Terms of Business* (Cambridge, Mass.: The Medieval Academy of America, 1934), Appendix III, pp. 392–404.

26. Frederic C. Lane, *Andrea Barbarigo, Merchant of Venice* 1418–1449 (New York: Octagon Books, 1967), p. 144.

27. Ibid., p. 146.

28. Edler, op.cit. p. 21.

29. The dating of coins was a rather late European practice. The earliest known dated coin, using Roman numerals, appeared in Aachen in 1373. Coins with dates given in Hindu-Arabic numerals appeared in Switzerland (1424), Germany (1448), Sweden (1478), France (1485), Italy (1534), Scotland (1549), and England (1551). Latin countries began dating coins later than their northern neighbors. A curious anomaly to the above chronology is a coin from the Norman Sicily of Roger II bearing the date AH533 (A.D. 1138) in Hindu-Arabic numerals. *See* Rodolfo Spahr, *Le Monete Siciliane* (Zürich: Graz, 1976), p. 152. The author wishes to express his thanks to Dr. Alan M. Stahl, Associate Curator of Medieval Coins, The American Numismatic Society, for providing information on the dating of coins.

30. As determined by evidence given in G. F. Hill, *The development of Arabic Numerals in Europe* (Oxford: The Clarendon Press, 1915).

31. In the *Treviso,* zero is referred to as *cifra, cifer* and *nulla,* nothing. Menninger erroneously attributes the first use of the word *nulla* to an Italian arithmetic of 1484 (Menninger, op. cit., p. 403).

32. The numbers were developed on a six-digit basis and included: *byllion, tryllion, quadrillon, quyllion, sixlion, septyllion, ottyllion,* and *nonyllion.*

33. Florian Cajori, *A History of Mathematical Notations,* vol. I (La Salle, Ill.: The Open Court Publishing Co., 1928); J.W.L. Glaisher, "On the Early History of the Signs + and − and on the Early German Arithmeticians", *Messenger of Mathematics* (1921–22), 51: 1–148.

34. Fol. 3v.

35. Jackson, op. cit., p. 41.

36. Maximum Planudes (c.1340) and Fibonacci (1202) wrote the sum at the top of the problem.

37. 'Casting out sevens' was also a popular test of computational correctness.

38. In a true mathematical sense, the 'casting out of nines' is not a proof of the correctness of a computational result, but merely a check. A proof is a logical argument based on mathematical definitions and principles that asserts the correctness of a result without room for doubt. In contrast, a check is a quick and easy procedure used to affirm the correctness of an obtained answer. A check lacks the certainty of a proof. The 'casting out of nines' can indicate a mistake has been made, but it cannot insure a correct result has been achieved. For example, consider the problem:

$$\begin{array}{r} 25 \\ 86 \\ 92 \\ \hline 203 \end{array} \qquad \begin{array}{rcl} 25 & = & 7 \pmod 9 \\ 86 & = & 5 \pmod 9 \\ 92 & = & 2 \pmod 9 \\ \hline 203 & = & \overline{5} \pmod 9 \end{array}$$

The 'casting out of nines' in the right-hand computation indicates that the answer, 203, is not incorrect, but it does not guarantee the answer is correct. Addition results of 500, 95, 113, etc., would also supply the same residue, 5, when the nines are cast out of the sum.

39. Karpinski believes that the practice is carried over from the format followed in abacus arithmetics, since the abacus only allows for the addition of two numbers at one time (Karpinski, op. cit., p. 102). Fibonacci was one of the first authors to depart from this custom.

40. Both the Chinese and the Hindus were familiar with negative numbers, and Fibonacci in his *Flos* (c.1225) acknowledged a negative number to mean a loss rather than a gain, but this interpretation of negative numbers dealt with roots of algebraic equations. In Western arithmetic, negative numbers were not yet fully understood.

41. The first use of such a bar appears in Prosdocimo de' Beldamandi, *Algorithmus de integris* (1410).

42. The other two methods were the 'borrowing and repaying scheme' as used by Borghi (1484) and the 'plan of simple borrowing'. For details of these methods, *see* Smith op. cit., 2: pp. 99–100.

43. Fol. 11v.

Notes to Chapter 4

1. Fol. 14r.
2. Robert Steele, "The Earliest Arithmetic in English" (London: Early English Text Society, 1922), p. 21.
3. Rhabdas (1341) indicates the technique was taught to him by Palamedes. Paul Tannery, *Notices et extraits des manuscrits de la Bibliothèque nationale,* 32: 167.
4. It is found in the *Kholasat al-hisab* of Beha Eddin (c.1600).
5. For a more detailed discussion of these methods, *see* Jackson, op. cit., pp. 62–64 or Smith, op cit., 2: 107–123.
6. As translated in Menninger, op. cit., pp. 429–30.
7. Fol. 15v.
8. The author employs the spelling *scachiero* throughout his work: however, the accepted Venetian spelling of the word was *scachieri,* after the term for chessboard, *sarchero.*
9. Smith, op. cit., 2: 110.
10. The journey westward is traced out in Smith, op. cit., 2: 115–16.
11. John Napier, *Rabdologiae, sev numerationes per virgulas libri duo* (Edinburgh, 1617).

Notes to Chapter 5

1. Fol. 22v.
2. Lucas Pacioli, *Suma de arithmetica, geometria, proportioni et proportionalita . . .* (Venice: 1494), Fol. 32v.
3. The term *lauanzo* apparently evolved from *l'avenzo,* meaning a surplus, or in a business context, a profit.
4. Hylles, *Arithmeticke* (1600), Fol. 37.
5. "*. . . regulae quae a sudantibus abaustis vix intelleguntur*", translated in Menninger, op. cit., p. 327.
6. Also discussed in Menninger, op. cit., p. 327.
7. Smith, op. cit., 2:134.

8. Nicolo Tartaglia, *Tutte l'opera d'arithmetica del famosissimo Nicoló Tartaglia* (Venice: 1592), Fol. 53.

9. For a discussion of Venetian maritime experience, *see* Frederic Lane, *Venice and History* (Baltimore: The Johns Hopkins Press, 1966), "Ships and Shipping", pp. 143–263.

10. As quoted in Smith, op. cit., 2:137.

11. *See* Lam Lay Yong, "On the Chinese Origin of the Galley Method of Arithmetical Division", *British Journal of the History of Science* (1966), 3:66–69.

12. As quoted in Smith, op. cit., 2:140.

13. For a more complete discussion of the various methods of division, *see* Jackson, op. cit., pp. 69–77.

14. Philipi Calanderi, *Philippi Calandri ad nobilem et studiosum julianum laurentü medicem de arithmethrica opusculum* (Florence: 1491).

15. American secondary mathematics books of the 1920s employed this term.

16. The first formal presentation of a theory of decimal fractions was made by Simon Stevin in his treatise *De Thiende* [*La Disme*], 1585.

Notes to Chapter 6

1. Sixty-two pages.

2. In the Ahmes, or Rhind, Papyrus, problem 72 asks for the number of loaves of bread of "strength" 45 which are equivalent to 100 loaves of "strength" 10. *See* A. B. Chace, et al., *The Rhind Mathematical Papyrus* (Washington, D.C.: The National Council of Teachers of Mathematics, 1979), reprint of 1927–29 edition; *Tshui fen* chapter of *Chiu chang suan shu*. *See* Frank Swetz and T. I. Kao, *Was Pythagoras Chinese?* (University Park, Penn.: The Pennsylvania State University Press, 1977), pp. 18–19.

3. H. T. Colebrooke, *Algebra with Arithmetic and Mensuration from the Sanskrit of Brahmagupta and Bhascara* (Calcutta, 1927), p. 283.

4. Fol. 30r.

5. *See* Smith, op. cit., 2:484–488.

6. James Hodder, *Arithmetick or that Necessary Art Made Most Easy* (London: 1672) tenth ed., p. 87.

7. Humphrey Baker, *The Well Spring of Science . . .* (London, 1568).

8. Fol. 31v.

9. Fol. 33r.

10. Fol. 34r.

11. Fol. 40v.

12. Fol. 36v–37r.

13. Fol. 43v.

14. T. Hylles, *Arithmeticks* (London, 1600), fol. 135.

15. Paolo Gerardi, *Libro di ragioni* (1328), as translated by Van Egmond, op. cit., p. 251.

16. Ibid.

17. Fol. 30v.

18. In following the theory formulated by G. H. F. Nesselmann in 1842, the evolution of algebra consists of three stages: a rhetorical stage where problems and their solutions are completely written out in words; a syncopated stage in which abbreviations are adopted for frequently-used terms and operations; and a symbolic stage in which a complete mathematical shorthand is employed. European algebra at this time is just entering the syncopated stage.

19. Pacioli, *Summa* (1494), fol. 67r, informs his readers that what is commonly called "la regola de la cosa" may also be called algebra, "ouer algebra e amucabala." Helmreich, *Rechenbüch* (1561), fol. 2 states that what the Italians call *Delacosa,* the Germans refer to as *Regula Cos* or algebra. Rudolff's algebra book of 1525 was entitled *Die Coss.*

20. *See* page from Benjamin Dearborn, *The Pupils' Guide* (Boston, 1782) reproduced in Karpinski, op. cit., p. 148.

21. Robert S. Lopez, "Stars and Spices: The Earliest Italian Manual of Commercial Practice", in *Economy, Society and Government in Medieval Italy,* David Herlihy, Robert Lopez and Vsevold Slessarev (eds.) (Kent, Ohio: The Kent State University Press, 1969), pp. 35–42.

22. Fol. 45r.

23. D. E. Smith, *Rara Arithmetica* (Boston: Ginn and Co., 1908), p. 440; Raymond de Roover, *Business, Banking and Economic Thought in Late Medieval and Early*

Modern Europe (Chicago: University of Chicago Press, 1974), p. 77.

24. For a complete discussion of Venetian business institutions, *see* Federic C. Lane, *Venice: A Maritime Republic* (Baltimore, Md.: The Johns Hopkins University Press, 1973), pp. 136–146; Edler, op. cit., p. 333.

25. *Liber abaci, Liber algorismi,* as given in B. Boncompagni, *Trattati d'Arithmetica* (Rome, 1857), 2:111.

26. Fol. 46v.

27. As quoted in Natalie Zemon Davis, "Sixteenth Century French Arithmetics on the Business Life", *Journal of the History of Ideas* (1960), 21: 18–48.

28. The strategies of medieval and renaissance bartering are discussed more fully in Van Egmond, op. cit., pp. 185–188. He notes that commercial arithmetics are a singular source of information on the practice of European mercantile barter.

29. *Ground of Artes* (1558).

30. Fol. 52v.

31. Fol. 54v.

32. Kou-ku, problem 14. *See* Frank J. Swetz and T. I. Kao, *Was Pythagoras Chinese? An Examination of Right Triangle Theory in Ancient China* (University Park, Pa.: The Pennsylvania State University Press, 1977), p. 46.

33. Fol. 55v.

34. Charlemagne established a court school for bright youths. The puzzles were probably intended to be used in their training.

35. This problem can also be found in the *Chiu chang suan shu*.

36. Fol. 56r.

37. As quoted in Robert S. Lopez, "Stars and Spices: The Earliest Italian Manual of Commercial Practice", *Economy, Society and Government in Medieval Italy,* David Herlihy, Robert Lopez and Vsevold Slessarev (eds.) (Kent, Ohio: The Kent State University Press, 1969), pp. 35–42.

38. Ibid., p. 41.

39. *pratica della mercatura*.

40. Translated by Van Egmond, op. cit., p. 117.

41. Fols. 57v–59r.

42. *See* George Zinberg, *Jewish Calendar Mystery Dispelled* (New York: Vantage Press, 1963).

43. A more detailed account of this dispute is given in P. W. Wilson, *Romance of the Calendar* (New York: W. W. Norton, 1937).

44. Named after the 5th-century B.C. Athenian astronomer Meton who first discovered it. A more modern method of computing the date of Easter was devised by the German mathematician Karl Friedrich Gauss in the early part of the nineteenth century. For a description of Gauss's method, *see* R. F. Graesser, "The Date of Easter", *School Science and Mathematics* (1952), 52: 371–372.

45. Bede incorporated this information in his compute. *See* entry on Bede in *Dictionary of Scientific Biography,* Charles C. Gillespie (ed.) (New York: Charles Scribner's Sons, 1970), 1: 564–566.

46. Fol. 58r. At this period of history, there existed no mechanical device with sufficient accuracy to indicate the passing of a second or a point. The "second" as a time measure did not appear until the late sixteenth century. It is apparent that the *puncto* was a unit derived for computational convenience. Its use was probably regional and extremely limited. However, the fact that a *puncto* is based on a factor of three rather than ten or sixty, the bases for established mathematical systems of measurement, remains interesting and unexplained. The author is indebted to Stillman Drake of the University of Toronto for sharing his thoughts with him on the occurrence of *puncto* as a unit of time measure in the *Treviso Arithmetic.*

Notes to Chapter 7

1. Pompeo Molment, *Venice: Its Individual Growth From the Earliest Beginning to the Fall of the Republic,* 2 vols. (Chicago: A. C. McClurg and Company, 1906), 1:126.

2. List given in Jackson, op. cit., p. 147.

3. Gino Luzzatto, *An Economic History of Italy from the Fall of the Roman Empire to the Beginning of the Sixteenth Century* (London: Routledge and Kegan Paul, 1961), p. 116.

4. Venice was Europe's major silk-producing city at this time.

5. Lane, op. cit., p. 313.

6. A detailed discussion of the Italian wool industry is given in Edler, op. cit., pp. 409–19. *See also* Raymond de Roover, "A Florentine Firm of Cloth Manufacturers", *Speculum* (1941), 16:3–33.

7. The major industry of the city of Venice, itself, was an arsenal, a complex in which ships for its maritime fleet would be built and fitted. Since the arsenal cannot be considered a commercial venture, it has not been discussed above.

8. A history of spices is given in John W. Parry, *The Story of Spices* (New York: Chemical Publishing Company, 1953).

9. Harry A. Miskimin, *The Economy of Early Renaissance Europe* 1300–1460 (Englewood Cliffs, N.J.: Prentice-Hall, 1969), p. 127.

10. As quoted by Daniel J. Boorstin, *The Discoverers* (New York: Random House, 1983), p. 195.

11. Miskimin, op. cit., p. 127.

12. Samuel Perkins, "The Magic of Saffron", *Silver Kris* (1980), 5:29–31.

13. As an example of this feminine ingenuity consider the technique used by Venetian women to tint their hair; tresses were dipped in a mixture of sulphur, egg shell, and orange peel and then bathed in the sunlight to effect a color change. Titian's "Venus of Urbino" (1538) bears the hair coloring. John Evelyn, a 17th-century English diarist, in commenting on the hairstyles of Venetian women, noted that "They weare very long crisped hair of several strakes and colours, which they artifically make so, by washing their heads in pisse."

14. William Shakespeare, *The Winter's Tale,* act 4, sc. 3.

15. The Merchant's Tale. *See* Nevill Coghill, *The Canterbury Tales,* (Harmondsworth: Penguin Books, 1951), p. 395.

16. Warren Dawson, *A Leechbook or Collection of Medical Recipes of the Fifteenth Century* (M.S. #136, Medical Society of London). (London: Macmillan & Co., 1934).

17. John Payne (translator), *The Decameron of Giovanni Boccaccio* (New York: Horace Liveright, 1925), p. 11. During the time of the Black Death citizens of Venice wore a protective covering which included a face mask. This mask had a beak-like protrusion over the nose and mouth which contained spices to 'filter' the air.

18. As given in Lane, op. cit., p. 333.

19. Burke, op. cit., p. 18.

20. Luzzatto, op. cit., p. 131.

21. Some other prices extracted from arithmetic books are given in David Eugene Smith, "Mathematical Problems in Relation to the History of Economics and Commerce", *American Mathematical Monthly* (1918), 24:221–23.

22. John Ball, the 'Mad Priest of Kent' who was executed in 1381 for causing unrest. Quoted in Parry, op. cit., p. 82.

23. Luzzatto, op. cit., p. 114.

24. Molment, op. cit., 2:6.

25. Translated in *The Renaissance: Maker of Modern Man,* National Geographic Society (Washington, D.C.: National Geographic Book Service, 1970), p. 93.

26. Smith, "Mathematical Problems . . .", op. cit.

27. Fol. 43v.

28. As related in Lane, op. cit., p. 332.

29. Carlo M. Cipolla, *Money, Prices, and Civilization in the Mediterranean World* (New York: Gordian Press, Inc., 1967); Elgin Groseclose, *Money and Man: A Survey of Monetary Experience* (New York: Frederick Ungar Publishing Company, 1961).

30. A detailed discussion of the ducat and its influence is given in Herbert E. Ives, *The Venetian Gold Ducat and its Imitations* (New York: The American Numismatic Society, 1954).

31. Luzzatto, op. cit., p. 126.

32. Molment, op. cit., 2:160.

33. A complete listing of all exchange rates given by Tartaglia can be found in Jackson, op. cit., p. 147.

34. W. K. Ferguson, "Recent Trends in the Economic Historigraphy of the Renaissance", *Studies in the Renaissance* (1960), 7:19–26.

35. Lane, *The Collected Papers,* op. cit., p. 50.

36. A 'letter' or 'bill of exchange' was a document prepared by a merchant house for one of its distant agents to supply the bearer with an indicated sum of money or goods. This innovation, the forerunner of a modern bank draft, eliminated the physical transport of large sums of money. The term 'endorse', from the Latin *dorsum,* back, was used to describe the notations made on the back of such documents.

37. Gail, op. cit., p. 91.

38. Gail, op. cit., p. 91.

39. For a more complete discussion of the restrictions regarding usury, *see* B. N. Nelson, "The Usurer and the Merchant Prince: Italian Businessmen and the Ecclesiastical Law of Restitution, 1100–1550", *Journal of Economic History* (1947) 7 (Sup.): 104–23; John F. McGovern, "The Rise of New Economic Attitudes—Economic Humanism, Economic Nationalism—During the Later Middle Ages and the Renaissance, A.D. 1200–1550", *Traditio* (1970), 26:217–253.

40. Kirshner, op. cit., p. 123.

41. *See* Abbott Payson Usher, *The Early History of Deposit Banking in Mediterranean Europe* (New York: Russell and Russell, 1943).

42. Jeannin, op. cit., p. 38.

43. Incident related in S. Carl Hirsch, *Meter Means Measure* (New York: Viking Press, 1973).

44. Lopez, "Stars and Spices", op. cit., p. 40.

45. Lopez, "Stars and Spices", op. cit., p. 40.

46. Further information on Italian weights and measures in the fifteenth and sixteenth centuries is given in Edler, op. cit., pp. 317–22.

47. Vogel gives different equivalents. *See* Kurt Vogel, *Ein Italienisches Rechenbuch aus dem 14 Jahrhundert* (Munich: Deutschen Museums für die Geschichte der Naturwissenschaften und der Technik, 1977).

48. A thorough study of the evolution of weights and measures in England is given in Ronald Zupko, *A Dictionary of English Weights and Measures from Anglo-Saxon Times to the Nineteenth Century* (Madison, Wisc.: The University of Wisconsin Press, 1968).

49. Many early arithmetic books had no formal titles.

50. Fol. 1r.

51. Fols. 15v; 25r; 62r.

52. Translated from the Latin in Menninger, op. cit., p. 434.

53. Statement attributed to Augustine (400). Douglas M. Campbell, *The Whole Craft of Number* (Boston: Prindle, Weber & Schmidt, Inc., 1976).

54. Douglas McMurtrie, op. cit., p. 283.

55. Van Egmond, op. cit., p. 322.

56. As translated in R. Emmett Taylor, *No Royal Road: Luca Pacioli and his Times* (Chapel Hill: The University of North Carolina Press, 1942), p. 63.

57. A more complete discussion of *practica della mercatura* is given in Lopez, "Stars and Spices", op. cit.

58. Paul L. Rose, *The Italian Renaissance of Mathematics* (Geneva: Librairie Droz, 1975), p. 1.

59. Arthur O. Garder, "The History of Mathematics as a Part of the History of Mankind", *The Mathematics Teacher* (1968), pp. 73–80.

60. *See* the discussion given by Lynn Thorndike, "Renaissance or Prenaissance?", *Journal of the History of Ideas* (1943), 4:65–74.

61. George Sarton, "Science in the Renaissance", in *The Civilization of the Renaissance,* James W. Thompson et al. (Chicago: University of Chicago Press, 1929), pp. 75–95; comments by Jacob Bronowski are also worth noting; *see Magic, Science and Civilization* (New York: Columbia University Press, 1978).

62. *See* G. F. Hill, *The Development of Arabic Numerals in Europe* (Oxford: The Clarendon Press, 1915).

63. G. A. Miller, "On the History of Common Fractions", *School Science and Mathematics* (1931), 31:138–145.

64. Calculation by percentage is said to have originated in Venice about 1224. *See* H. Gordon Selfridge, *The Romance of Commerce* (London: John Lane, The Bodley Head, Ltd., 1923), p. 56.

65. *See* Jeannin, op. cit., particularly "The Merchant Mentality and the Scientific Spirit", pp. 108–112.

66. Van Egmond, op. cit., presents a detailed discussion of this genre referring to the works in question as "abaci". Van Egmond has also prepared a catalogue of extant copies of commercial arithmetics of this period, *Practical Mathematics in the Italian Renaissance: A Catalog of Italian Abacus Manuscripts and Printed Books to*

1600 (Florence: Instituto e Museo di Storia della Scienza, 1980). While the merchant profession had the most extensive literature devoted to its mathematical needs, other professions also had their own mathematical literature. Geometric form and proportion were discussed in texts for use by artists; *see,* for example, Margaret Daly Davis, *Piero della Francesca's Mathematical Treatises* (Revenna: Longo, 1977). Arithmetic texts were also compiled for use by the military, e.g., *Arithmeticall Militare Treatise, named Stratioticos,* Leonard and Thomas Digges (1572), *L'art d'arythmetique contenant toute dimention . . . tant pour l'art militaire que pour autres calculations,* Claude de Boissiere (1554).

67. Fol. 1r.

68. Taylor, op. cit., p. 63.

69. *See* David E. Smith, "The Influence of the Mathematical Works of the Fifteenth Century upon those of Later Times", *Papers of the Bibliographical Society of America* (1932), 26:143–171; George Sarton, *Appreciation of Ancient and Medieval Science During the Renaissance* (New York: A. S. Barnes & Company, 1961).

70. The theory of the Renaissance as a time of economic depression has been developed by Robert Lopez. *See* his "Hard Times and Investment in Culture", *The Renaissance: A Symposium* (New York: The Metropolitan Museum of Art, 1953); *The Commercial Revolution of the Middle Ages,* 950–1350 (Englewood Cliffs, N.J.: Prentice-Hall, 1971).

71. A complete discussion of German commercial arithmetic is given by Hugo Grosse, *Historische Rechenbücher des 16 und 17 Jahrhundertes und die Entwicklung ihrer Grundgedanken bis zur Neuzeit* (Leipzig: Dürr, 1901).

72. Frederic C. Lane, "An Anonymous Merchant of the Fourteenth Century: How to Succeed in Business while Trying", *The Collected Papers,* op. cit., pp. 53–58.

73. *See* Hans L. Wussing, "European Mathematics During the Evolutionary Period of Early Capitalistic Conditions", *Organon* (1967), 4:89–93.

74. J. A. G. Shirk, "Contributions of Commerce to Mathematics," *The Mathematics Teacher* (1939), 32:203–208.

75. Fol. 62r.

Index